Elastic wave propagation in transversely isotropic media

Monographs and textbooks on mechanics of solids and fluids
editor-in-chief: G. Æ. Oravas

Mechanics of elastic and inelastic solids
editor: S. Nemat-Nasser

Books published under this series are:

1. G.M.L. Gladwell
   Contact problems in the classical theory of elasticity
2. G. Wempner
   Mechanics of solids with applications to thin bodies
3. T. Mura
   Micromechanics of defects in solids
4. R.G. Payton
   Elastic wave propagation in transversely isotropic media
5. S. Nemat-Nasser, H. Abé, S. Hirakawa (Eds.)
   Hydraulic fracturing and geothermal energy

Elastic wave propagation in transversely isotropic media

Robert G. Payton

Department of Mathematics and Computer Science
Adelphi University, Garden City, N.Y. U.S.A.

1983 **MARTINUS NIJHOFF PUBLISHERS**
a member of the KLUWER ACADEMIC PUBLISHERS GROUP
THE HAGUE / BOSTON / LANCASTER

Distributors

for the United States and Canada: Kluwer Boston, Inc., 190 Old Derby Street, Hingham, MA 02043, USA
for all other countries: Kluwer Academic Publishers Group, Distribution Center, P.O.Box 322, 3300 AH Dordrecht, The Netherlands

Library of Congress Cataloging in Publication Data

```
Payton, Robert G.
   Elastic wave propagation in transversely isotropic
media.

   (Mechanics of elastic and inelastic solids ; 4)
(Monographs and textbooks on mechanics of solids and
fluids)
   Includes index.
   1. Elastic waves. 2. Wave-motion, Theory of.
I. Title. II. Series. III. Series: Monographs and
textbooks on mechanics of solids and fluids.
QA935.P34  1983      531'.3823        83-6289
ISBN 978-94-009-6868-4
```

ISBN 978-94-009-6868-4   ISBN 978-94-009-6866-0 (eBook)
DOI 10.1007/978-94-009-6866-0

Copyright

IV

Ter nagedachtenis aan mijn moeder, Ada Wright Payton (1896–1980), voor mijn vader Paul Payton en voor mijn geliefde vrouw Marlene.

So teach us to number our days, that we may apply our hearts unto wisdom.

psalm 90, v. 12.

Table of contents

Contents

Preface

In this monograph I record those parts of the theory of transverse isotropic elastic wave propagation which lend themselves to an exact treatment, within the framework of linear theory. Emphasis is placed on transient wave motion problems in two- and three-dimensional unbounded and semibounded solids for which explicit results can be obtained, without resort to approximate methods of integration.The mathematical techniques used, many of which appear here in book form for the first time, will be of interest to applied mathematicians, engeneers and scientists whose specialty includes crystal acoustics, crystal optics, magnetogasdynamics, dislocation theory, seismology and fibre wound composites.

My interest in the subject of anisotropic wave motion had its origin in the study of small deformations superposed on large deformations of elastic solids. By varying the initial stretch in a homogeneously deformed solid, it is possible to synthesize anisotropic materials whose elastic parameters vary continuously. The range of the parameter variation is limited by stability considerations in the case of small deformations superposed on large deformation problems and (what is essentially the same thing) by the notion of hyperbolicity (solids whose parameters allow wave motion) for anisotropic solids. The full implication of hyperbolicity for anisotropic elastic solids has never been previously examined, and even now the constraints which it imposes on the elasticity constants have only been examined for the class of transversely isotropic (hexagonal crystals) materials. In chapter one these constraints are discussed along with their relation to the more familiar constraints of positive definiteness.

Chapter two treats the shape and geometric features of the wave front caused by a point source. This is a beautiful subject on which significant work has been done by Musgrave (86). The wave front lends itself to an elegant solution in the form of an envelope to a plane wave solution of the nonlinear first order characteristic partial differential equation of the system. One of the high points of this chapter is an explicit set of conditions which, when imposed on the elasticity parameters, insure the existence of inflection points on the slowness curve. These conditions are the key to the classification of the wave front shape and to other propagational phenomena.

It is not generally known that an explicit solution may be found for the two-dimensional anisotropic elastic wave Green's tensor. But such is indeed the case and a detailed presentation of this remarkable result is presented in chapter three. This elegant solution is made possible by employing an integration technique due to Burridge (31). In this connection a full discussion of lacunas (gaps in the wave field behind the leading front) is given. Further explicit results are obtained along the symmetry axis for the three-dimensional solid.

The study of transient elastic waves in an elastic half-space began in 1904 when

Lamb analyzed the surface motion of a two-dimensional isotropic solid. Later Kraut (58), using the method of Cagniard, extended Lamb's results to transversely isotropic materials, in particular to the crystal Beryl. In chapter four the two-dimensional surface motion problem of a point loaded half-space is completely solved for the full range of hexagonal crystals. Finally the monograph ends in chapter five with some very recent research concerned with the epicentral-axis motion of a (surface) point loaded, three-dimensional half-space. Some intriguing results (for Zinc) caused by the head wave are herein given.

Musgrave's 1970 book 'Crystal Acoustics' [20] was the first '... to describe in detail the conservative mechanics of waves in crystals whether regarded as continua or as structures of discrete particles.' This book went far toward exploring the underlying physics associated with elastic waves in crystals. During the past decade several excellent books on the subject of linear, isotropic elastic wave motion have appeared in the English language. These books, by Achenbach [87], Eringen and Suhubi [88], Miklowitz [89] and Hudson [90] are written from the viewpoint of the applied mathematician. It has been my intention to write a monograph on transversely isotropic elastic waves, very much in the spirit of the above mentioned books. Emphasis has been placed on closed form solutions (no asymptotics!) to an interesting and important class of problems in continuum mechanics.

My thanks are due to Professor S. Nemat-Nasser for encouragement during the preparation of this monograph.

Garden City, New York   R.G. Payton
1983

Oh that my words were now written! oh that they were printed in a book!

Job 19, v. 23.

Basic equations

1. The Linearized Equations of Motion of an Anisotropic Elastic Solid

The linearized set of partial differential equations which govern the motion of a homogeneous anisotropic elastic solid are,

$$\text{equations of force equilibrium:} \quad \frac{\partial \sigma_{ji}}{\partial x_j} + \rho f_i = \rho \frac{\partial^2 u_i}{\partial t^2}, \tag{1.1.1}$$

$$\text{equations of moment equilibrium:} \quad \sigma_{ij} = \sigma_{ji}, \tag{1.1.2}$$

$$\text{strain-displacement relations:} \quad e_{ij} = \frac{1}{2}\left(\frac{\partial u_i}{\partial x_j} + \frac{\partial u_j}{\partial x_i}\right), \tag{1.1.3}$$

$$\text{stress-strain relations:} \quad \sigma_{ij} = c_{ijkl} e_{kl}. \tag{1.1.4}$$

The above equations, which hold within the interior of the solid, are written with respect to a fixed Cartesian reference frame $Ox_1 x_2 x_3$, with t denoting time. σ_{ij} and e_{ij} denote the Cartesian components of the stress and strain tensors respectively, while u_i denotes the components of the displacement vector. The components of the body force vector, per unit mass, are f_i and ρ is the constant mass density of the solid. The fourth order tensor of the elasticities c_{ijkl} satisfy the (Green) symmetry conditions

$$c_{ijkl} = c_{jikl} = c_{klij}, \tag{1.1.5}$$

so that there are but 21 independent constants needed to describe the stress-strain relations for a general anisotropic elastic solid. Finally, in order to formulate stress boundary conditions, the relation between the surface traction vector T_i acting on a surface element with an outward drawn unit normal vector n_i, and the stress tensor σ_{ij} is

$$T_i = \sigma_{ij} n_j.$$

In eqns. (1.1.1)–(1.1.5) the indicies i and j range over the integers 1, 2 and 3, with repeated indicies summed from 1 to 3.

Equations (1.1.1)–(1.1.4) constitute a set of 15 equations for the 15 unknowns σ_{ij}, e_{ij} and u_i. Since the equations are linear with constant coefficients, it is an easy matter to reduce this set to 3 equations for the 3 components of the displacement vector u_i,

$$c_{ijkl} \frac{\partial^2 u_k}{\partial x_l \partial x_j} + \rho f_i = \rho \frac{\partial^2 u_i}{\partial t^2}. \tag{1.1.6}$$

In the case of an isotropic solid, the c_{ijkl} tensor is constrained to satisfy

$$c_{ijkl} = \lambda \delta_{ij} \delta_{kl} + \mu (\delta_{ik} \delta_{jl} + \delta_{il} \delta_{jk}),$$ (1.1.7)

where λ and μ are the Lamé material constants. Insertion of (1.1.7) into eqn. (1.1.6) gives the famous equations named for Navier

$$(\lambda + \mu) \frac{\partial^2 u_k}{\partial x_k \partial x_i} + \mu \frac{\partial^2 u_i}{\partial x_k \partial x_k} + \rho f_i = \rho \frac{\partial^2 u_i}{\partial t^2}.$$ (1.1.8)

The equations of this section require considerable space if carefully derived. However this hardly seems necessary here since their derivation is given in books devoted to the theory to elasticity, e.g., Green and Zerna [1], Pearson [2] and Sokolnikoff [3].

2. The Effect on the Equations of Motion of a Coordinate Rotation

As will be shown in the next section, the equations of motion for a transversely isotropic solid take on their most attractive form when displayed with respect to a coordinate frame in which one axis coincides with the axis of material symmetry. However, due to the orientation of the loading or bounding surface, this frame may not be the most natural one in which to formulate the problem. Consider then a second right-handed Cartesian coordinate frame $Ox_1' x_2' x_3'$ which is obtained from the first frame $Ox_1 x_2 x_3$, by means of a rotation. The rotation is specified by the 9 numbers α_{ij} where

$$\alpha_{ij} = \text{cosine (angle between } Ox_i' \text{ and } Ox_j\text{).}$$ (1.2.1)

Then, according to the rules of Cartesian tensor transformation (see, e.g., Pearson [2]) the equations of motion in the primed coordinate system have the form

$$c_{ijkl}' \frac{\partial^2 u_k'}{\partial x_i' \partial x_j'} + \rho f_i' = \rho \frac{\partial^2 u_i'}{\partial t^2},$$ (1.2.2)

where

$$c_{ijkl}' = \alpha_{ip} \alpha_{jq} \alpha_{kr} \alpha_{ls} c_{pqrs}.$$ (1.2.3)

Although eqns. (1.1.6) and (1.2.2) appear to have the same form, relation (1.2.3) shows that in fact there may be a considerable difference in the number of terms involved when the equations are written out in full. In this respect the equations of motion for an isotropic elastic solid are very special since (by definition of isotropy)

$$c_{ijkl}' = c_{ijkl}.$$ (1.2.4)

3. The Elasticities for a Transversely Isotropic Solid

Because of the various symmetries in the components of c_{ijkl} caused by (1.1.5), crystallographers have introduced a shortened matrix notation to describe the stress-strain relation for an anisotropic solid. In this notation, pairs of subscripts involving the numbers 1, 2 and 3 are replaced by the numbers 1, 2, 3, 4, 5 and 6 as follows

$$(11) \leftrightarrow 1, (22) \leftrightarrow 2, (33) \leftrightarrow 3, (23) = (32) \leftrightarrow 4, (31) = (13) \leftrightarrow 5 \text{ and}$$

$$(12) = (21) \leftrightarrow 6. \tag{1.3.1}$$

So that, e.g.,

$$c_{1232} = c_{64}.$$

Using this notation, Fedorov [4] shows that the stress-strain relations (1.1.4) for a transversely isotropic material, whose axis of symmetry coincides with the x_3-axis, is given by

$$
\begin{bmatrix} \sigma_{11} \\ \sigma_{22} \\ \sigma_{33} \\ \sigma_{23} \\ \sigma_{31} \\ \sigma_{12} \end{bmatrix}
=
\begin{bmatrix}
c_{11} & c_{12} & c_{13} & 0 & 0 & 0 \\
c_{12} & c_{11} & c_{13} & 0 & 0 & 0 \\
c_{13} & c_{13} & c_{33} & 0 & 0 & 0 \\
0 & 0 & 0 & c_{44} & 0 & 0 \\
0 & 0 & 0 & 0 & c_{44} & 0 \\
0 & 0 & 0 & 0 & 0 & (c_{11} - c_{12})/2
\end{bmatrix}
\begin{bmatrix} e_{11} \\ e_{22} \\ e_{33} \\ 2e_{23} \\ 2e_{31} \\ 2e_{12} \end{bmatrix}
\tag{1.3.2}
$$

If a primed coordinate system is being used in which the x_3'-axis is not parallel to the symmetry axis, then eqns. (1.3.1), (1.3.2) together with (1.2.2) and (1.2.3) will determine the equations of motion. It should be noted, however, that the elasticity tensor c_{ijkl} remains invariant with respect to coordinate frame rotations about the symmetry axis. This justifies the description of such media as transversely isotropic.

An isotropic solid is a special case of a transversely isotropic solid for which

$$c_{11} = c_{33} = \lambda + 2\mu, \qquad c_{12} = c_{13} = \lambda \quad \text{and} \quad c_{44} = \mu. \tag{1.3.3}$$

Consequently there are but two material constants (not counting the density ρ)

Table 1. Elasticity Constants for Some Hexagonal Crystals.

Crystal	c_{11}	c_{12}	c_{13}	c_{33}	c_{44}
Apatite	16.7	1.31	6.6	14.0	6.63
Beryllium	29.2	2.67	1.4	33.6	16.2
Beryl	28.2	9.94	6.95	24.8	6.86
Cadmium	11.6	4.23	4.14	5.10	1.95
Cobalt	30.7	16.5	10.3	35.8	7.55
Ice (257K)	1.35	0.65	0.52	1.45	0.317
Hafnium	18.1	7.7	6.6	19.7	5.57
Magnesium	5.92	2.57	2.14	6.14	1.64
Rhenium	61.2	27.0	20.6	68.3	16.2
Titanium	16.2	9.2	6.9	18.1	4.67
Thallium	4.08	3.54	2.9	5.28	0.726
Yttrium	7.79	2.92	2.0	7.69	2.431
Zinc	16.5	3.1	5.0	6.2	3.96

All the tabulated numbers should be multiplied by 10^{11} dyne/cm^2.

needed to specify an isotropic solid, whereas · a transversely isotropic material requires the specification of five material constants c_{11}, c_{12}, c_{13}, c_{33} and c_{44} for its description. Some typical values of these constants for 13 hexagonal crystals have been extracted from Hearman [5] and presented in Table 1.

4. The Constrains on the $C_{\alpha\beta}$'s of Positive Definiteness

The elasticities c_{ijkl} are said to be positive definite when

$$c_{ijkl} a_{ij} a_{kl} > 0 \tag{1.4.1}$$

for arbitrary nonzero tensors a_{ij}. Physically this constraint corresponds to the requirement that the strain energy density

$$\epsilon = \tfrac{1}{2} c_{ijkl} e_{ij} e_{kl} \tag{1.4.2}$$

must remain positive since this energy must be minimal in a state of stable equilibrium. In terms of the matrix notation of section 3, the positive definite condition (1.4.1) becomes

$$c_{\alpha\beta} a_\alpha a_\beta > 0 \quad \text{for} \quad \alpha,\beta = 1, 2, \ldots, 6, \tag{1.4.3}$$

for arbitrary nonzero vectors a_α. From the theory of linear algebra (see Noble [6]), the necessary and sufficient conditions on the elements of $c_{\alpha\beta}$ in order that the quadratic form (1.4.3) be positive are that the six determinants

$$c_{11}, \begin{vmatrix} c_{11} & c_{12} \\ c_{21} & c_{22} \end{vmatrix}, \begin{vmatrix} c_{11} & c_{12} & c_{13} \\ c_{21} & c_{22} & c_{23} \\ c_{31} & c_{32} & c_{33} \end{vmatrix}, \ldots, \det c_{\alpha\beta}, \tag{1.4.4}$$

remain positive. Using the form of the matrix $c_{\alpha\beta}$ from (1.3.2), appropriate to a transversely isotropic solid, in (1.4.4) gives

$$c_{11} > |c_{12}|, \quad (c_{11} + c_{12})c_{33} > 2c_{13}^2 \quad \text{and} \quad c_{44} > 0 \tag{1.4.5}$$

The corresponding constraints on an isotropic solid are

$$(3\lambda + 2\mu) > 0 \quad \text{and} \quad \mu > 0. \tag{1.4.6}$$

A nice discussion of the positive definite constraints for all crystal classes, not just transversely isotropic, is given in Fedorov [4].

5. The Constraints on the $C_{\alpha\beta}$'s of Strong Ellipticity

If the equations of motion (1.1.6) are to remain hyperbolic, then the homogeneous set of equations

$$c_{ijkl} \frac{\partial^2 u_k}{\partial x_l \partial x_j} = \rho \frac{\partial^2 u_i}{\partial t^2} \tag{1.5.1}$$

should admit plane wave solutions of the form

$$u_i = g_i(\hat{\omega} \cdot \mathbf{x} - Vt), \tag{1.5.2}$$

for an arbitrary unit vector $\hat{\omega}$. Here $\hat{\omega}$ is the normal vector to the plane wave moving with speed V

$$\hat{\omega} \cdot \mathbf{x} - Vt = \text{constant}. \tag{1.5.3}$$

Substitution of (1.5.2) into (1.5.1) gives

$$(c_{ijkl}\omega_l\,\omega_j - \rho V^2\delta_{ik})g_k'' = 0. \tag{1.5.4}$$

In order that nontrivial solutions to (1.5.4) exist

$$\det(c_{ijkl}\omega_l\omega_j - \rho V^2\delta_{ik}) = 0. \tag{1.5.5}$$

The definition of hyperbolicity for the set of partial differential equations (1.1.6) (or (1.5.1)) as given by John [7] or Gel'fand and Shilov [8] requires that the sixth degree equation (1.5.5) in V must have 6 real roots for any (real) values of ω_1, ω_2 and ω_3 such that $\omega_1^2 + \omega_2^2 + \omega_3^2 = 1$. The constraints placed on the elasticities by the condition of hyperbolicity is the same as that given by Hadamard for infinitesimal stability of a solid experiencing a small deformation superimposed on a finite deformation. The implications of Hadamard's condition are discussed at some length by Truesdell and Noll [9], who point out that non-propagating ($V = 0$) singular surfaces are not excluded by the condition of stability, while by the strong ellipticity condition they are.

In contrast to hyperbolicity, the conditions of strong ellipticity imply that the V^2 roots of (1.5.5) are positive so that zero propagation speeds are ruled out. In this monograph the condition of strong ellipticity will be used since this condition seems most natural for wave propagation problems.

Knops and Payne [10] give the definition of strong ellipticity for the system (1.1.6) as

$$c_{ijkl}\omega_l\,\omega_j a_i a_k > 0, \tag{1.5.6}$$

for all nonzero vectors ω_i and a_i. Now the V^2 roots of (1.5.5) are the eigenvalues (to within a multiplicative constant ρ) of the matrix

$$B_{ik} = c_{ijkl}\omega_l\,\omega_j. \tag{1.5.7}$$

But the matrix B_{ik} is symmetric since

$$B_{ik} = c_{ijkl}\omega_l\omega_j = c_{klij}\omega_l\omega_j = c_{kjil}\omega_j\omega_l = B_{ki}. \tag{1.5.8}$$

According to Courant [11], positive eigenvalues of the matrix B_{ik} are necessary and sufficient conditions for

$$B_{ik}a_i a_k > 0 \tag{1.5.9}$$

for all $a_i \not\equiv 0$, which is just condition (1.5.6). Thus, *strong ellipticity implies real, nonzero propagation speeds*. The constraints imposed on the elasticities by strong ellipticity will now be derived.

In operator notation, eqns. (1.1.6) may be written as,

$$L_{ij}[u_j] = -\rho f_i \qquad (1.5.10)$$

where

$$L_{11} = c_{11}\frac{\partial^2}{\partial x^2} + \tfrac{1}{2}(c_{11} - c_{12})\frac{\partial^2}{\partial y^2} + c_{44}\frac{\partial^2}{\partial z^2} - \rho\frac{\partial^2}{\partial t^2},$$

$$L_{12} = L_{21} = \tfrac{1}{2}(c_{11} + c_{12})\frac{\partial^2}{\partial x \partial y},$$

$$L_{13} = L_{31} = (c_{13} + c_{14})\frac{\partial^2}{\partial x \partial z},$$

$$L_{22} = \tfrac{1}{2}(c_{11} - c_{12})\frac{\partial^2}{\partial x^2} + c_{11}\frac{\partial^2}{\partial y^2} + c_{44}\frac{\partial^2}{\partial z^2} - \rho\frac{\partial^2}{\partial t^2},$$

$$L_{23} = L_{32} = (c_{13} + c_{44})\frac{\partial^2}{\partial y \partial z}, \quad \text{and}$$

$$L_{33} = c_{44}\frac{\partial^2}{\partial x^2} + c_{44}\frac{\partial^2}{\partial y^2} + c_{33}\frac{\partial^2}{\partial z^2} - \rho\frac{\partial^2}{\partial t^2}. \qquad (1.5.11)$$

In (1.5.11) and henceforth, the coordinates x_1, x_2, x_3 have been replaced by the more familiar x, y, z so that the z-axis is now the axis of material symmetry.

Since the partial differential equations (1.5.10) have constant coefficients, they may be uncoupled to give a single sixth order equation for u_1, u_2 or u_3. The differential operator on the lefthand side of these three equations is the same in all cases, namely

$$F\left(\frac{\partial}{\partial x}, \frac{\partial}{\partial y}, \frac{\partial}{\partial z}, \frac{\partial}{\partial t}\right) \equiv \det(L_{ij}), \qquad (1.5.12)$$

where

$$F\left(\frac{\partial}{\partial x}, \frac{\partial}{\partial y}, \frac{\partial}{\partial z}, \frac{\partial}{\partial t}\right)$$

$$= \left[c_{44}\frac{\partial^2}{\partial z^2} + \tfrac{1}{2}(c_{11} - c_{12})\left(\frac{\partial^2}{\partial x^2} + \frac{\partial^2}{\partial y^2}\right) - \rho\frac{\partial^2}{\partial t^2}\right]\left[c_{44}c_{33}\frac{\partial^4}{\partial z^4}\right.$$

$$+ \left\{c_{44}^2 + c_{11}c_{33} - (c_{13} + c_{44})^2\right\}\frac{\partial^2}{\partial z^2}\left(\frac{\partial^2}{\partial x^2} + \frac{\partial^2}{\partial y^2}\right)$$

$$+ c_{44}c_{11}\left(\frac{\partial^2}{\partial x^2} + \frac{\partial^2}{\partial y^2}\right)^2 - \rho(c_{33} + c_{44})\frac{\partial^4}{\partial z^2 \partial t^2}$$

$$\left. - \rho(c_{11} + c_{44})\frac{\partial^2}{\partial t^2}\left(\frac{\partial^2}{\partial x^2} + \frac{\partial^2}{\partial y^2}\right) + \rho^2\frac{\partial^4}{\partial t^4}\right]. \qquad (1.5.13)$$

The fact that the sixth degree operator F can be factored into the product of a second degree operator and a fourth degree operator was apparently first noticed by Christoffel (1877). This, together with the transverse isotropy simplification (i.e., the appearance of x and y derivatives only in the form $\partial^2/\partial x^2 + \partial^2/\partial y^2$), is one of the main reasons why some problems in wave propagation for such materials are amenable to analysis. In the case of isotropy, (1.5.13) becomes

$$F = \left[\mu\left(\frac{\partial^2}{\partial z^2} + \frac{\partial^2}{\partial x^2} + \frac{\partial^2}{\partial y^2}\right) - \rho\frac{\partial^2}{\partial t^2}\right]^2$$

$$\times \left[(\lambda + 2\mu)\left(\frac{\partial^2}{\partial z^2} + \frac{\partial^2}{\partial x^2} + \frac{\partial^2}{\partial y^2}\right) - \rho\frac{\partial^2}{\partial t^2}\right]. \tag{1.5.14}$$

The condition of strong ellipticity says that the sixth degree algebraic equation in V

$$F(\omega_1, \omega_2, \omega_3, -V) = 0 \tag{1.5.15}$$

must have six real, nonzero roots for all unit vectors $\hat{\omega}$. Take $\hat{\omega} = (0, 0, 1)$ then

$$F(0, 0, 1, -V) = [c_{44} - \rho V^2]\,[c_{44}c_{33} - \rho(c_{33} + c_{44})V^2 + \rho^2 V^4]$$

$$= 0, \tag{1.5.16}$$

so

$$\rho V^2 = c_{44}, \qquad \rho V^2 = c_{33}, \qquad \rho V^2 = c_{44}$$

hence

$$c_{33} > 0 \quad \text{and} \quad c_{44} > 0. \tag{1.5.17}$$

Next take $\hat{\omega} = (\cos\theta, \sin\theta, 0)$ with $0 \leqslant \theta < 2\pi$, then

$$F(\cos\theta, \sin\theta, 0, -V) = [\tfrac{1}{2}(c_{11} - c_{12}) - \rho V^2]$$

$$\times [c_{44}c_{11} - \rho(c_{11} + c_{44})V^2 + \rho^2 V^4] = 0, \tag{1.5.18}$$

so

$$\rho V^2 = \tfrac{1}{2}(c_{11} - c_{12}), \qquad \rho V^2 = c_{11}, \rho V^2 = c_{44},$$

hence

$$c_{11} > 0, \qquad c_{44} > 0 \quad \text{and} \quad \tfrac{1}{2}(c_{11} - c_{12}) > 0. \tag{1.5.19}$$

Since $c_{44} > 0$, divide F by c_{44}^3 and write

$$\frac{1}{c_{44}^3}F\left(\frac{\partial}{\partial x}, \frac{\partial}{\partial y}, \frac{\partial}{\partial z}, \frac{\partial}{\partial t}\right) \equiv L\left(\frac{\partial}{\partial x}, \frac{\partial}{\partial y}, \frac{\partial}{\partial z}, \frac{\partial}{\partial \tau}\right)Q\left(\frac{\partial}{\partial x}, \frac{\partial}{\partial y}, \frac{\partial}{\partial z}, \frac{\partial}{\partial \tau}\right),$$

where

$$L\left(\frac{\partial}{\partial x}, \frac{\partial}{\partial y}, \frac{\partial}{\partial z}, \frac{\partial}{\partial \tau}\right) = \frac{\partial^2}{\partial z^2} + \delta\left(\frac{\partial^2}{\partial x^2} + \frac{\partial^2}{\partial y^2}\right) - \frac{\partial^2}{\partial \tau^2} \tag{1.5.20}$$

$$Q\left(\frac{\partial}{\partial x}, \frac{\partial}{\partial y}, \frac{\partial}{\partial z}, \frac{\partial}{\partial \tau}\right) = \alpha \frac{\partial^4}{\partial z^4} + \gamma \frac{\partial^2}{\partial z^2}\left(\frac{\partial^2}{\partial x^2} + \frac{\partial^2}{\partial y^2}\right)$$

$$+ \beta \left(\frac{\partial^2}{\partial x^2} + \frac{\partial^2}{\partial y^2}\right)^2 - (\alpha + 1)\frac{\partial^4}{\partial z^2 \partial \tau^2}$$

$$- (\beta + 1)\frac{\partial^2}{\partial \tau^2}\left(\frac{\partial^2}{\partial x^2} + \frac{\partial^2}{\partial y^2}\right) + \frac{\partial^4}{\partial \tau^4}. \qquad (1.5.21)$$

Here $\alpha = c_{33}/c_{44}$, $\beta = c_{11}/c_{44}$, $\gamma = 1 + \alpha\beta - (c_{13}/c_{44} + 1)^2$ and

$$\delta = \tfrac{1}{2}(\beta - c_{12}/c_{44}). \qquad (1.5.22)$$

The four dimensionless parameters α, β, γ and δ have replaced the five elasticities c_{11}, c_{12}, c_{13}, c_{33} and c_{44}. Also

$$\tau = \sqrt{\frac{c_{44}}{\rho}}\, t \qquad (1.5.23)$$

is the new time coordinate (with the physical dimension of length). The operators L and Q are very important in the study of wave propagation problems and will appear in many different forms throughout this monograph.

At this stage $\alpha > 0$, $\beta > 0$ and $\delta > 0$ are necessary conditions for positive V^2 roots of (1.5.15). Now take $\hat{\omega} = (\sin\theta\cos\phi, \sin\theta\sin\phi, \cos\theta)$ with $0 \leqslant \theta \leqslant \pi$, $0 \leqslant \phi < 2\pi$, then $L(\omega_1, \omega_2, \omega_3, -V) = \cos^2\theta + \delta\sin^2\theta - V^2$ so that

$$\delta > 0 \qquad (1.5.24)$$

is a necessary and sufficient condition for $V^2 > 0$ root of $L = 0$. Now

$$Q(\omega_1, \omega_2, \omega_3, -V) = A(\theta) - V^2 B(\theta) + V^4 \qquad (1.5.25)$$

where

$$A(\theta) = \alpha\cos^4\theta + \gamma\cos^2\theta\sin^2\theta + \beta\sin^4\theta, \qquad (1.5.26)$$

and

$$B(\theta) = (\alpha + 1)\cos^2\theta + (\beta + 1)\sin^2\theta. \qquad (1.5.27)$$

Since θ appears in A and B only in the form $\cos^2\theta$ and $\sin^2\theta$ it is only necessary to consider the θ interval $0 \leqslant \theta \leqslant \pi/2$. In order to complete the strong ellipticity requirement on Q, the question is now raised: what conditions must be imposed on α, β and γ in order that the quadratic equation in V^2

$$A - BV^2 + V^4 = 0, \qquad (1.5.28)$$

have positive roots? If $Q = 0$ is to have four real (nonzero) roots then

(i) $A(\theta) > 0$ since both V^2 roots must be positive, $(1.5.29)$

(ii) $B(\theta) > 0$ since both V^2 roots must be positive, $(1.5.30)$

(iii) $(B^2 - 4A) \geqslant 0$ since both V^2 roots must be real.   (1.5.31)

Consequence of $A(\theta) > 0$ for $0 \leqslant \theta \leqslant \pi/2$.

If $\theta = 0$ then $A(0) = \alpha > 0$. If $\theta \neq 0$ then

$$A(\theta)/\sin^4 \theta = \alpha \cot^4 \theta + \gamma \cot^2 \theta + \beta \tag{1.5.32}$$

and as θ goes from 0 to $\pi/2$, $\cot \theta$ goes from $+\infty$ to zero. Call

$$g(x) = \alpha x^2 + \gamma x + \beta. \tag{1.5.33}$$

Then the condition $A(\theta) > 0$ implies that $g(x) > 0$ for $0 \leqslant x < \infty$, i.e., $g(x)$ cannot have a positive root. But the roots of $g(x)$ are

$$x_{\pm} = \frac{-\gamma \pm \sqrt{\gamma^2 - 4\alpha\beta}}{2\alpha}. \tag{1.5.34}$$

Three possibilities for $g(x)$ must be considered.

1. $(\gamma^2 - 4\alpha\beta) < 0$ then both $x_{\pm}$ are complex so that $g(x) > 0$ for all $x \geqslant 0$, using $g(0) = \beta > 0$.
2. $(\gamma^2 - 4\alpha\beta) = 0$ and $\gamma > 0$, then both $x_{\pm}$ are negative so that $g(x) > 0$ for all $x \geqslant 0$.
3. $(\gamma^2 - 4\alpha\beta) > 0$ with $\gamma > 0$, then both $x_{\pm}$ are negative so that $g(x) > 0$ for all $x \geqslant 0$.

These conditions may be combined to give

$$A(\theta) > 0 \text{ provided } \alpha > 0, \beta > 0 \text{ and } \gamma > -2\sqrt{\alpha\beta}. \tag{1.5.35}$$

Consequence of $B(\theta) > 0$ for $0 \leqslant \theta \leqslant \pi/2$.

Since $B(\theta) = (\alpha + 1)\cos^2 \theta + (\beta + 1)\sin^2\theta$, this condition is automatically satisfied when $\alpha > 0$ and $\beta > 0$.

Consequence of $B^2(\theta) - 4A(\theta) > 0$.

(The situation for $B^2 - 4A = 0$ will be mentioned below separately.) Now

$$B^2 - 4A = a \cos^4\theta + c \cos^2 \theta \sin^2 \theta + b \sin^4 \theta,$$

and

$$(B^2 - 4A)/\sin^4 \theta = a \cot^4 \theta + c \cot^2\theta + b, \tag{1.5.36}$$

where

$$a = (\alpha - 1)^2, \qquad c = 2[(\alpha + 1)(\beta + 1) - 2\gamma], \qquad b = (\beta - 1)^2. \tag{1.5.37}$$

From comparing (1.5.37) with (1.5.32) it is immediately seen that

$$(B^2 - 4A) > 0 \text{ implies } a > 0, b > 0 \text{ and } c > -2\sqrt{ab}. \tag{1.5.38}$$

9

The condition $c > -2\sqrt{ab}$ is the same as

$$2(\alpha + 1)(\beta + 1) - 4\gamma > -2\sqrt{(\alpha - 1)^2(\beta - 1)^2}. \tag{1.5.39}$$

Four cases must now be distinguished.

1. $\alpha < 1, \beta < 1$ so that $(\alpha + 1)(\beta + 1) - 2\gamma > -(1 - \alpha)(1 - \beta)$
   hence $\gamma < (\alpha\beta + 1)$, $\tag{1.5.40}$

2. $\alpha < 1, \beta > 1$ so that $(\alpha + 1)(\beta + 1) - 2\gamma > -(1 - \alpha)(\beta - 1)$
   hence $\gamma < (\alpha + \beta)$, $\tag{1.5.41}$

3. $\alpha > 1, \beta < 1$ so that $(\alpha + 1)(\beta + 1) - 2\gamma > -(\alpha - 1)(1 - \beta)$
   hence $\gamma < (\alpha + \beta)$, $\tag{1.5.42}$

4. $\alpha > 1, \beta > 1$ so that $(\alpha + 1)(\beta + 1) - 2\gamma > -(\alpha - 1)(\beta - 1)$
   hence $\gamma < (\alpha\beta + 1)$. $\tag{1.5.43}$

But from the definition of γ in (1.5.22) it is clear that

$$\gamma \leqslant (1 + \alpha\beta), \tag{1.5.44}$$

and in cases 2. and 3. above $(1 + \alpha\beta) < (\alpha + \beta)$ since $(1 - \alpha)(1 - \beta) < 0$. Hence for these two cases the upper bound for γ is $(1 + \alpha\beta)$ also.

Remark. When the four V-roots of $Q(\omega_1, \omega_2, \omega_3 - V) = 0$ are real and distinct, then the partial differential operator Q of (1.5.21) is said to be *strictly hyperbolic*. The conditions for strict hyperbolicity are $A(\theta) > 0, B(\theta) > 0$ and $B^2(\theta) - 4A(\theta) > 0$, and from eqns. (1.5.40)–(1.5.44) they imply case

1) $0 < \alpha < 1$, $\quad 0 < \beta < 1$, $\quad -2\sqrt{\alpha\beta} < \gamma < (\alpha\beta + 1)$ $\tag{1.5.45}$

2) $0 < \alpha < 1$, $\quad \beta > 1$, $\quad -2\sqrt{\alpha\beta} < \gamma \leqslant (\alpha\beta + 1)$ $\tag{1.5.46}$

3) $\alpha > 1$, $\quad 0 < \beta < 1$, $\quad -2\sqrt{\alpha\beta} < \gamma \leqslant (\alpha\beta + 1)$, $\tag{1.5.47}$

4) $\alpha > 1$, $\quad \beta > 1$, $\quad -2\sqrt{\alpha\beta} < \gamma < (\alpha\beta + 1)$. $\tag{1.5.48}$

Remark. The condition $\gamma \leqslant (1 + \alpha\beta)$ for cases 2) and 3) is required for the reality of the elasticities, not from the mathematical definition of hyperbolicity. The equation $Q[u] = f$ is a perfectly valid strictly hyperbolic equation for $(1 + \alpha\beta) \leqslant \gamma < (\alpha + \beta)$ under cases 2) and 3) and may have a physical interpretation for some other continuum theory.

Consequence of $B^2(\theta) - 4A(\theta) = 0$.

Under the above conditions, there will be repeated V-roots of $Q(\omega_1, \omega_2, \omega_3, -V) = 0$. The analysis is straight forward and has been treated by Payton [12]. Six additional cases are introduced. Case

5) $0 < \alpha < 1$, $\quad 0 < \beta < 1$, $\quad \gamma = (\alpha\beta + 1)$ $\tag{1.5.49}$

6) $\alpha > 1$, $\quad \beta > 1$, $\quad \gamma = (\alpha\beta + 1)$ $\tag{1.5.50}$

7) $\alpha = 1$, $\quad \beta \neq 1$, $\quad -2\sqrt{\beta} < \gamma \leqslant (1 + \beta)$ $\tag{1.5.51}$

8) $\alpha \neq 1,$ $\beta = 1,$ $-2\sqrt{\alpha} < \gamma \leqslant (1 + \alpha)$ (1.5.52)

9) $\alpha = \beta = 1,$ $-2 < \gamma < 2$ (1.5.53)

10) $\alpha = \beta = 1,$ $\gamma = 2.$ (1.5.54)

In summary: the partial differential operators L and Q of (1.5.20) and (1.5.21) remain strongly elliptic or hyperbolic with nonzero propagation speeds for a real material provided $\delta > 0$ and α, β and γ fall into cases 1)–10) above. The corresponding constraints on the elasticities are

$$c_{11} > 0, \qquad c_{11} > c_{12}, \qquad c_{33} > 0, \qquad c_{44} > 0$$

and

$$|c_{13} + c_{44}| < (\sqrt{c_{11}c_{33}} + c_{44}). \qquad (1.5.55)$$

The corresponding constraints for an isotropic solid ($\alpha = \beta = \gamma/2$) are

$$(\lambda + 2\mu) > 0 \quad \text{and} \quad \mu > 0. \qquad (1.5.56)$$

Comparing the constraints of positive definiteness (1.4.5) and those for strong ellipticity (1.5.55) (respectively (1.4.6) and (1.5.56) for an isotropic solid) it is apparent that the condition of strong ellipticity is less restrictive on the elasticities than those of positive definiteness. This observation can, of course, be seen directly from the definitions (1.4.1) and (1.5.6).

The distinctions between the various cases 1)–10) of this section will be of importance in Chapter 2 when discussing the wave front shape caused by point and line sources. The material in this section is based on Payton [13] and [14].

6. The Uncoupled Equations of Motion in Two Dimensions

In this section the two dimensional equations of motion appropriate to a point impulsive source (a line source in three dimensions) will be given. The analysis of this problem, and the corresponding three dimensional problem, provides the major theme for this monograph.

Let the body force per unit mass be applied in the form

$$f_i = (0, a_2 \delta(y)\delta(z)\delta(t), a_3 \delta(y)\delta(z)\delta(t)), \qquad (1.6.1)$$

where δ denotes the Dirac delta. Physically the loading (1.6.1) corresponds to a uniform line load applied perpendicular to the symmetry axis (the z-axis) along (say) the x-axis. The constants a_2 and a_3 are polarization coefficients of the source. Further let (u, v, w), rather than (u_1, u_2, u_3), be the components of the displacement vector with respect to the frame $Oxyz$. Since the loading does not vary in the x-direction and $u \equiv 0$, the problem is one of plane strain. Writing

$$v = (\rho/c_{44})^{1/2}(a_2 v_2 + a_3 v_3) \quad \text{and} \quad w = (\rho/c_{44})^{1/2}(a_2 w_2 + a_3 w_3) \qquad (1.6.2)$$

and noting that the polarization coefficients a_2 and a_3 may be independently equated to zero, the equation of motion (1.1.6) may be uncoupled to give

$$Q[v_2] = -N[f], \qquad Q[v_3] = \kappa \frac{\partial^2}{\partial y \partial z}[f],$$

$$Q[w_2] = \kappa \frac{\partial^2}{\partial y \partial z}[f] \quad \text{and} \quad Q[w_3] = -K[f]. \tag{1.6.3}$$

In (1.6.3)

$$f = \delta(y)\delta(z)\delta(\tau) \quad \text{and} \quad \kappa = (1 + \alpha\beta - \gamma)^{1/2}. \tag{1.6.4}$$

The partial differential operators Q, N and K are given by

$$Q\left(\frac{\partial}{\partial y}, \frac{\partial}{\partial z}, \frac{\partial}{\partial \tau}\right) = \alpha \frac{\partial^4}{\partial z^4} + \gamma \frac{\partial^4}{\partial z^2 \partial y^2} + \beta \frac{\partial^4}{\partial y^4} - (\alpha + 1)\frac{\partial^4}{\partial z^2 \partial \tau^2}$$

$$- (\beta + 1)\frac{\partial^4}{\partial y^2 \partial \tau^2} + \frac{\partial^4}{\partial \tau^4}, \tag{1.6.5}$$

$$N\left(\frac{\partial}{\partial y}, \frac{\partial}{\partial z}, \frac{\partial}{\partial \tau}\right) = \alpha \frac{\partial^2}{\partial z^2} + \frac{\partial^2}{\partial y^2} - \frac{\partial^2}{\partial \tau^2}, \tag{1.6.6}$$

and

$$K\left(\frac{\partial}{\partial y}, \frac{\partial}{\partial z}, \frac{\partial}{\partial \tau}\right) = \frac{\partial^2}{\partial z^2} + \beta \frac{\partial^2}{\partial y^2} - \frac{\partial^2}{\partial \tau^2}. \tag{1.6.7}$$

In eqns. (1.6.3), $v_3(y, z, \tau)$ represents the y-displacement component due to a line load (uniform along the x-axis) polarized in the z-direction. The other displacement components may be similarly interpreted. The Q operator (1.6.5) is, of course, the same as (1.5.21) provided the x-derivatives are suppressed.

If the line load is not perpendicular to the axis of material symmetry, then the coordinate axes should be rotated to a new frame in which one coordinate axis again lies along the line load. The problem is still one of plain strain. But now, due to the rotated form of the equations of motion (1.2.2), the partial differential operators appearing on the left hand side of (1.6.3) will be of sixth rather than fourth order. In terms of the rotated coordinates, the operators L (which will now appear) and Q become more extended since they will now contain a richer variety of mixed derivatives (see section 3.8).

7. The Uncoupled Equations of Motion in Three Dimensions

The basic three dimensional problem to be treated in Chapters 2 and 3 concerns a body force of the form

$$f_i = (a_1, a_2, a_3)\delta(x)\delta(y)\delta(z)H(t), \tag{1.7.1}$$

where $H(t)$ is the Heaviside unit step function in time. The loading (1.7.1) corresponds to a point source which is suddenly applied at the origin and thereafter maintained for all time. Writing

$$u = \rho(a_1 u_1 + a_2 u_2 + a_3 u_3)/c_{44}, \qquad v = \rho(a_1 v_1 + a_2 v_2 + a_3 v_3)/c_{44}$$

and $\qquad\qquad\qquad\qquad\qquad\qquad\qquad\qquad\qquad\qquad\qquad\qquad$ (1.7.2)

$$w = \rho(a_1 w_1 + a_2 w_2 + a_3 w_3),$$

for the displacement vector (u, v, w), and noting that the polarization coefficients a_1, a_2 and a_3 may be independently equated to zero, gives for the uncoupled equations of motion (1.1.6)

$$LQ[u_1] = -\left(Q + \frac{\partial^2}{\partial x^2} M\right)[f], \qquad LQ[u_2] = -\left(\frac{\partial^2}{\partial x \partial y} M\right)[f],$$

$$Q[u_3] = \kappa \left(\frac{\partial^2}{\partial x \partial z}\right)[f], \qquad LQ[v_1] = -\left(\frac{\partial^2}{\partial x \partial y} M\right)[f],$$

$$\qquad\qquad\qquad\qquad\qquad\qquad\qquad\qquad\qquad\qquad\qquad (1.7.3)$$

$$LQ[v_2] = -\left(Q + \frac{\partial^2}{\partial y^2} M\right)[f], \qquad Q[v_3] = \kappa \left(\frac{\partial^2}{\partial y \partial z}\right)[f],$$

$$Q[w_1] = \kappa \left(\frac{\partial^2}{\partial x \partial z}\right)[f], \qquad Q[w_2] = \kappa \left(\frac{\partial^2}{\partial y \partial z}\right)[f],$$

and

$$Q[w_3] = -K[f].$$

In eqns. (1.7.3)

$$f = \delta(x)\delta(y)\delta(z)H(t) \qquad\qquad\qquad\qquad\qquad (1.7.4)$$

and the operators M and K are given by

$$M\left(\frac{\partial}{\partial x}, \frac{\partial}{\partial y}, \frac{\partial}{\partial z}, \frac{\partial}{\partial \tau}\right) = (1 + \alpha\delta - \gamma)\frac{\partial^2}{\partial z^2} + (\delta - \beta)\left(\frac{\partial^2}{\partial x^2} + \frac{\partial^2}{\partial y^2}\right)$$

$$-(\delta - \beta)\frac{\partial^2}{\partial \tau^2},$$

and

$$K\left(\frac{\partial}{\partial x}, \frac{\partial}{\partial y}, \frac{\partial}{\partial z}, \frac{\partial}{\partial \tau}\right) = \frac{\partial^2}{\partial z^2} + \beta\left(\frac{\partial^2}{\partial x^2} + \frac{\partial^2}{\partial y^2}\right) - \frac{\partial^2}{\partial \tau^2}. \qquad (1.7.5)$$

The operators L and Q, for the three dimensional case, have been previously defined in eqns. (1.5.20) and (1.5.21).

13

8. Some Other Transversely Isotropic Continuum Theories

Although the main attention of this monograph will be directed toward transversely isotropic elastic media (and in particular hexagonal crystals), the analysis is by no means limited to such problems. Several examples from other continuum theories which fit the form of the above equations, will now be cited.

(i) Small deformations superposed on finite deformations.

In a two dimensional problem, the Q operator of (1.6.5) has parameters (Payton [13], Green and Zerna [1])

$$\alpha = (c_{66} + \tau^{11})/(c_{11} + 2\tau^{11}), \qquad \beta = (c_{66} + \tau^{22})/(c_{22} + 2\tau^{22})$$

and

$$\gamma = 1 + \alpha\beta - \frac{(c_{12} + \tau^{11} + c_{66})^2}{(c_{11} + 2\tau^{11})(c_{22} + 2\tau^{22})} \tag{1.8.1}$$

where the c_{ij}'s depend on the homogeneous stretch parameters through the strain energy density function. Here τ^{11} and τ^{22} measure the biaxial state of stress at infinity.

(ii) Magneto-hydrodynamic waves.

In a three dimensional problem (Lighthill [15]) the Q operator of (1.5.21) has parameters

$$\alpha = a_0^2/a_1^2, \qquad \beta = 0, \qquad \gamma = a_0^2/(a_0^2 + a_1^2) \tag{1.8.2}$$

where a_0 is the sound speed and a_1 is the speed at which Alfvén waves propagate.

(iii) Crystal optics in uniaxial crystals.

In a two dimensional problem (Born and Wolf [16]) the Q operator of (1.6.5) has parameters

$$\alpha = \sigma^2, \qquad \beta = 1, \qquad \gamma = \sigma^2 + 1, \tag{1.8.3}$$

where σ involves the optical constants of the crystal. In this case the Q operator can be factored to

$$Q = \left(\frac{\partial^2}{\partial t^2} - \frac{\partial^2}{\partial y^2} - \frac{\partial^2}{\partial z^2}\right)\left(\frac{\partial^2}{\partial t^2} - \frac{\partial^2}{\partial y^2} - \sigma^2 \frac{\partial^2}{\partial z^2}\right). \tag{1.8.4}$$

(iv) Composite materials.

In a two dimensional problem (Moon [17]) the Q operator of (1.6.5) has the same formal definitions for α, β and γ as (1.5.22) except that now the 'crystal constants' depend on the properties and geometric arrangement of the composite constituents. In this case the parameters α, β and γ may be varied continuously by changing the fiber lay-up angle.

Wave front shape caused by a point source in unbounded media

The major problem treated in this monograph is that of constructing the free space Green's function for a transversely isotropic solid. In this chapter the theory of characteristics of hyperbolic partial differential equations will be used to analyze the wave front shape. In Chapter 3 the more difficult problem of constructing the solution behind the wave front will be studied. Major references for this chapter are volume II of Courant and Hilbert [18], Kline and Kay [19] and Musgrave [20] who did much of the early work on wave fronts in anisotropic elastic media.

Part I. TWO SPACE DIMENSIONS

1. The Characteristic Partial Differential Equation for the Wave Front

For an initial value problem in which the disturbance f is localized at the origin $(y = z = 0)$, a wave front will form at $\tau = 0$ and propagate away from this point with increasing time. The geometry of this front, which separates the undisturbed region ahead of the front from the disturbed region behind the front, is the main subject of this chapter.

Let $\phi(y, z, \tau) = 0$ be the equation of the wave front. Introduce new independent variables ϕ, ϕ_1 and ϕ_2 in place of y, z, and τ with the understanding that jumps in the displacement derivatives, e.g., $\dfrac{\partial^2 v}{\partial \phi^2}$ and $\dfrac{\partial^2 w}{\partial \phi^2}$ may occur across the *wave front*

$$W: \phi(y, z. \tau) = 0, \tag{2.1.1}$$

but that other derivatives $\dfrac{\partial^2 v}{\partial \phi \partial \phi_1}$, $\dfrac{\partial^2 v}{\partial \phi_1^2}$, $\dfrac{\partial^2 v}{\partial \phi_1 \partial \phi_2}$, $\dfrac{\partial^2 v}{\partial \phi_2^2}$, etc. are continuous there. By the chain rule

$$\frac{\partial v}{\partial y} = \frac{\partial v}{\partial \phi} \frac{\partial \phi}{\partial y} + \frac{\partial v}{\partial \phi_1} \frac{\partial \phi_1}{\partial y} + \frac{\partial v}{\partial \phi_2} \frac{\partial \phi_2}{\partial y}, \tag{2.1.2}$$

and

$$\frac{\partial^2 v}{\partial y^2} = \frac{\partial^2 v}{\partial \phi^2} \left(\frac{\partial \phi}{\partial y} \right)^2 + \dots \tag{2.1.3}$$

where the dots denote terms which are continuous across W. The two dimensional version of (1.5.10) now become (with $u_1 = 0, u_2 = v, u_3 = w, \dfrac{\partial}{\partial x_1} = 0, x_2 = y$ and $x_3 = z$)

$$(\beta\phi_y^2 + \phi_z^2 - \phi_\tau^2)\frac{\partial^2 v}{\partial\phi^2} + \kappa\phi_y\phi_z\frac{\partial^2 w}{\partial\phi^2} + \ldots = -\frac{\rho}{c_{44}}a_2 f,$$

$$\kappa\phi_y\phi_z\frac{\partial^2 v}{\partial\phi^2} + (\phi_y^2 + \alpha\phi_z^2 - \phi_\tau^2)\frac{\partial^2 w}{\partial\phi^2} + \ldots = -\frac{\rho}{c_{44}}a_3 f.$$

(2.1.4)

Here the abbreviated notation for partial derivatives

$$\frac{\partial\phi}{\partial y} = \phi_y \text{ etc.}$$

is used. If the set of equations (2.1.4) is evaluated at two neighboring points A and B on opposite sides of the wave front, and then these two points are allowed to approach a common point C on W, the difference between the two sets gives

$$(\beta\phi_y^2 + \phi_z^2 - \phi_\tau^2)\left[\frac{\partial^2 v}{\partial\phi^2}\right] + \kappa\phi_y\phi_z\left[\frac{\partial^2 w}{\partial\phi^2}\right] = 0,$$

$$\kappa\phi_y\phi_z\left[\frac{\partial^2 v}{\partial\phi^2}\right] + (\phi_y^2 + \alpha\phi_z^2 - \phi_\tau^2)\left[\frac{\partial^2 w}{\partial\phi^2}\right] = 0,$$

(2.1.5)

where $\left[\dfrac{\partial^2 v}{\partial\phi^2}\right]$ and $\left[\dfrac{\partial^2 w}{\partial\phi^2}\right]$ are the jumps in the second derivatives at C.

If these jumps are in fact nonzero then

$$(\beta\phi_y^2 + \phi_z^2 - \phi_\tau^2)(\phi_y^2 + \alpha\phi_z^2 - \phi_\tau^2) - \kappa^2\phi_y^2\phi_z^2 = 0. \tag{2.1.6}$$

By comparing (2.1.6) with (1.6.5) and noting that C is an arbitrary point on W, it is seen that (2.1.6) is equivalent to

$$Q(\phi_y, \phi_z, \phi_\tau) = 0 \quad \text{on} \quad \phi(y, z, \tau) = 0. \tag{2.1.7}$$

Equation (2.1.7) gives the characteristic partial differential equation for the wave front W associated with the two dimensional Green's function problem formulated in section (1.6).

According to the theory of partial differential equations, discontinuities in the second and higher derivatives of (1.5.10) will propagate on characteristic surfaces. (Indeed, in the sense of distributions, discontinuities in first order derivatives as well as in the displacement itself will propagate on such surfaces.) Thus the wave front is a characteristic surface.

As noted by Kline and Kay [19], the first of eqns. (2.1.7) is not a true partial differential equation since it holds on the surface $\phi(y, z, \tau) = 0$. However, assuming $\phi_\tau \neq 0$, the equation $\phi = 0$ may be solved for τ, so that

$$\phi(y, z, \tau) = \psi(y, z) - \tau = 0. \tag{2.1.8}$$

Substitution of (2.1.8) into (2.1.7) now gives a true first order, nonlinear partial differential equation for ψ

$$Q(\psi_y, \psi_z, -1) = 0. \tag{2.1.9}$$

2. The Normal Curve

At a particular instant of time $\tau > 0$, the wave front curve

$$W: \psi(y, z) - \tau = 0, \tag{2.2.1}$$

has a normal vector with components $\psi_y \equiv q$ and $\psi_z \equiv p$ which satisfy the fourth degree algebraic equation for the *normal curve*

$$N: Q(q, p, -1) \equiv F(p, q) = 0. \tag{2.2.2}$$

Explicitly, from (1.6.5)

$$F(p, q) = \alpha p^4 + \gamma p^2 q^2 + \beta q^4 - (\alpha + 1)p^2 - (\beta + 1)q^2 + 1. \tag{2.2.3}$$

The shape of the normal curve (called by some authors the "slowness curve") will play a decisive role in the construction of the wave front. For this reason a detailed analysis of the geometry of the normal curve will be given in this and the next four sections.

Instead of working directly with the p, q variables, it is advantageous to use polar coordinates defined by

$$p = R \cos \theta \quad \text{and} \quad q = R \sin \theta \quad \text{with} \quad 0 \leqslant \theta < 2\pi. \tag{2.2.4}$$

Equation (2.2.2) for the normal curve now becomes

$$AR^4 - BR^2 + 1 = 0, \tag{2.2.5}$$

where the functions $A(\theta)$ and $B(\theta)$ have been previously defined in eqs. (1.5.26) and (1.5.27). Solving (2.2.5) for R gives

$$R_\pm(\theta) = \left[\frac{B(\theta) \pm \sqrt{B^2(\theta) - 4A(\theta)}}{2A(\theta)} \right]^{1/2} \tag{2.2.6}$$

Since $A(\theta)$ and $B(\theta)$ remain positive by (1.5.29) and (1.5.30) and assuming the conditions for strict hyperbolicity (these conditions will be relaxed in section 5 in favor of strong ellipticity) it follows that

$$R_+(\theta) > R_-(\theta).$$ (2.2.7)

Thus the normal curve consists of two branches, in parameters form

$$N_\pm : p = R_\pm(\theta)\cos\theta \quad \text{and} \quad q = R_\pm(\theta)\sin\theta,$$ (2.2.8)

each of which surrounds the origin. The branches are in the form of closed ovals (because of the periodicity of the trigonometric functions involved) one of which $(R = R_-(\theta))$ lies completely inside the other $(R = R_+(\theta))$. These two branches of N possess double symmetry about the p and q axes since p and q only appear in (2.2.3) in the form of p^2 and q^2. Consequently if the shape of the normal curve is known for $0 \leqslant \theta \leqslant \pi/2$, then by reflection its shape is known for $0 \leqslant \theta < 2\pi$.

Consider the straight line

$$L: q = mp + b$$ (2.2.9)

in the p,q plane. How many times does L intersect N? At such an intersection point the q and p values of L and N must agree, hence

$$F(p, mp + b) = 0,$$ (2.2.10)

which is a quartic in p with, at most, four real roots. Now L can only intersect N_- twice since intersection of N_- implies intersection of N_+ (twice). Therefore N_- *is convex*. However it may be possible for L to intersect N_+ four times (in which case, of course, it would not intersect N_-) so that N_+ *may or may not be convex*. The situation in which N_+ is not convex has a profound influence on the wave front curve.

The above argument, establishing the convexity of the inner branch of the normal curve, is due to Duff [21].

3. Bitangents and the Existence of Inflection Points on the Normal Curve

If the outer branch of the normal curve is not convex, then it will contain inflection points. It turns out, and this is not a-priori obvious at this stage, that the existence or nonexistence of inflection points on N_+ is the key to the classification of the normal and wave front curve shapes. At first sight it might appear that (2.2.3) could be equated to zero, q determined as a function of p and that the vanishing of the second derivative of this function would give the inflection points. In theory this is true, but actually the resulting algebraic equation can only be treated numerically. A slightly less ambitious approach will be used instead. All permissible (α, β, γ) combinations will be sought which permit inflection points on N_+.

A unified approach to the existence of inflection points will now be established by finding the conditions under which bitangent lines exist for N_+. A bitangent line is a line which is tangent to a curve at two points. Obviously no such line exists for a convex curve. On the other hand if N_+ is not convex, then for every bitangent there will be two inflection points. In this section bitangents will be sought which are either vertical ($p =$ constant) or horizontal ($q =$ constant). The more difficult situation in which the bitangent line crosses both the p and q axes will be treated in the next section.

If a bitangent line to N_+ has the equation

$$q = q_0 > 0, \tag{2.3.1}$$

then this necessarily implies that q, regarded as a function of p on N_+, has a minimum at $p = 0$ or $\theta = \pi/2$. From the equation of the normal curve (2.2.2), its slope is found by (implicit) differentiation

$$\frac{dq}{dp} = -\frac{p[2\alpha p^2 + \gamma q^2 - (\alpha + 1)]}{q[2\beta q^2 + \gamma p^2 - (\beta + 1)]}. \tag{2.3.2}$$

At $\theta = \pi/2$, eqn. (2.2.6) gives

$$R_\pm(\pi/2) = \left[\frac{(\beta + 1) \pm |\beta - 1|}{2\beta}\right]^{1/2}, \tag{2.3.3}$$

so that

$$[2\beta q^2 + \gamma p^2 - (\beta + 1)] = \pm(\beta - 1) \neq 0 \quad \text{since} \quad \beta \neq 1.$$

(The situation where $\beta = 1$, together with other cases in which the operator Q ceases to be strictly hyperbolic, will be treated in section 5). Hence

$$\frac{dq}{dp}\bigg|_{\theta = \pi/2} = 0, \tag{2.3.4}$$

on both branches of the normal curve. Furthermore, for the outer branch

$$\frac{d^2q}{dp^2}\bigg|_{\theta = \pi/2} = -\frac{\gamma R_+^2(\pi/2) - (\alpha + 1)}{R_+(\pi/2)[2\beta R_+^2(\pi/2) - (\beta + 1)]}. \tag{2.3.5}$$

As an example, if $\beta > 1$ then (2.3.3) gives $R_+(\pi/2) = 1$ so

$$\frac{d^2q}{dp^2} = -\frac{\gamma - (\alpha + 1)}{\beta - 1} \quad \text{at} \quad \theta = \pi/2, \tag{2.3.6}$$

and this will be positive provided

$$\gamma < (\alpha + 1). \tag{2.3.7}$$

Thus if

$$-2\sqrt{\alpha\beta} < \gamma < (\alpha + 1) \quad \text{and} \quad \beta > 1, \tag{2.3.8}$$

N_+ will have (at least) four inflection points, one in each of the four quadrants (from symmetry).

By generalizing the above analysis to include vertical bitangents, it is seen that the algebraic sign of

$$\left. \frac{d^2 q}{dp^2} \right|_{\theta = \pi/2} \quad \text{and} \quad \left. \frac{d^2 p}{dq^2} \right|_{\theta = 0} \quad \text{on} \quad N_+ \tag{2.3.9}$$

will determine the number of inflection points on N_+ associated with horizontal and vertical bitangents. From Salmon [22], it is known that a quartic curve can have at most eight real inflection points. Thus if both terms in (2.3.9) are negative, there will be no inflection points (to be precise, no horizontal or vertical bitangents); if one term is positive and one negative there will be four inflection points; and if both terms are positive, there will be eight inflection points.

For the cases treated in this section the normal curve will have a variety of possible shapes depending on the relative values of the three parameters α, β and γ. The curve shape can be classified according to the behaviour of N_+ at $\theta = 0$ and $\theta = \pi/2$. From symmetry this branch will behave the same way at $\theta = \pi$ and $\theta = 3\pi/2$ respectively. In Tables 2−5 of section 6, the symbols MIN have been used for minimum, CINFL for two inflection points coalescing and MAX for maximum. The point at $\theta = 0$ on N_+ is classified as a minimum if the curve rises to the right, etc. As far as the general shape of the normal curve goes, it was found that there are five distinct classes.

References for this section are Musgrave [20] and Payton [13].

4. Bitangents Which Cross Both Coordinate Axes of the Normal Curve

If the (α, β, γ) combination is such that a bitangent line of the form

$$L: q = -mp + b, \quad m > 0, \quad b > 0 \tag{2.4.1}$$

exist then (from symmetry) this necessarily implies the existence of eight inflection points, two in each of the four quadrants of the p, q plane. But from the discussion of the last section it is known that N_+ can have at most eight inflection points. Hence it follows that if N_+ has a bitangent of the form (2.4.1) then N_+ will have a MAX point at both $\theta = 0$ and $\theta = \pi/2$. Anticipating the numbering scheme to be introduced in Tables 2−5 of section 6, bitangents of the form (2.4.1) may exist under cases 1.5, 1.8, 1.13, 2.5, 2.8, 2.13, 3.5, 3.8, 3.13, 4.5, 4.8 and 4.13.

The main idea of this section is to find, for a given (α, β) combination, the value of γ for which a bitangent ceases to exist (i.e. becomes a complex bitangent). This will occur when two inflection points coalesce. Let the first quadrant coordinates of such a point be (p_0, q_0). Recall the equation of the normal curve

$$F(p, q) = \alpha p^4 + \gamma p^2 q^2 + \beta q^4 - (\alpha + 1)p^2 - (\beta + 1)q^2 + 1 = 0. \qquad (2.4.2)$$

At the point (p_0, q_0), the polynomial

$$H(p) \equiv F(p, -mp + b) = a_4 p^4 + a_3 p^3 + a_2 p^2 + a_1 p + a_0 \qquad (2.4.3)$$

must satisfy the four conditions,

(i) $H(p_0)$ $= 0$ since the line L intersects N_+,
(ii) $H'(p_0)$ $= 0$ since the line L is tangent to N_+ at $p = p_0$,
(iii) $H''(p_0)$ $= 0$ since the line L is tangent to N_+ at a point where N_+ has an inflection
      point, and
(iv) $H'''(p_0)$ $= 0$ since the line L is tangent to N_+ at a point where N_+ has a double
      inflection point.

The above four conditions imply

$$H(p) = a_4(p - p_0)^4, \qquad (2.4.4)$$

which, when compared with eqn. (2.4.3), yields

$$a_3 = -4a_4 p_0, \qquad a_2 = 6a_4 p_0^2,$$
$$a_1 = -4a_4 p_0^3, \quad \text{and} \quad a_0 = a_4 p_0^4, \qquad (2.4.5)$$

where

$$a_0 = (\beta b^4 - \beta b^2 - b^2 + 1), \qquad a_1 = (-4mb^3 \beta + 2mb\beta + 2mb)$$
$$a_2 = (\gamma b^2 + 6m^2 b^2 \beta - \alpha - 1 - \beta m^2 - m^2), \qquad a_3 = (-2mb\gamma - 4m^3 b\beta)$$

and

$$a_4 = (\alpha + \gamma m^2 + \beta m^4).$$

The four nonlinear algebraic equations (2.4.5) serve to determine the unknowns p_0, m, b and place a constraint on the allowable (α, β, γ) combination which forces two inflection points to coalesce.

From eqn. (2.4.5) $a_3^2/a_1^2 = a_4/a_0$ so that

$$\frac{(\gamma + 2m^2 \beta)^2}{(1 + \beta - 2b^2 \beta)^2} = \frac{(\alpha + \gamma m^2 + \beta m^4)}{(\beta b^4 - \beta b^2 - b^2 + 1)}$$

which can be simplified to

$$\frac{(\gamma + 2m^2 \beta)^2}{(1 + \beta - 2b^2 \beta)^2} = \frac{\gamma^2 - 4\alpha\beta}{(\beta - 1)^2},$$

thus implying

$$\frac{a_0}{a_4} = \frac{(\beta - 1)^2}{\gamma^2 - 4\alpha\beta}.$$

But since $a_0/a_4 = p_0^4$, it follows that

$$p_0 = \left[\frac{(\beta - 1)^2}{\gamma^2 - 4\alpha\beta} \right]^{1/4}. \tag{2.4.6}$$

By an analogous procedure, the value of q_0 is found to be

$$q_0 = \left[\frac{(\alpha - 1)^2}{\gamma^2 - 4\alpha\beta} \right]^{1/4}. \tag{2.4.7}$$

Equations (2.4.6) and (2.4.7) give the coordinates (p_0, q_0) of a point on the normal curve where two inflection points coalesce. From symmetry there are three other such points $(p_0, -q_0)$, $(-p_0, q_0)$ and $(-p_0, -q_0)$. However, the specification of (p_0, q_0) is not really complete since the parameters α, β and γ are not independent. The (α, β, γ) constraint follows from the requirement that the point (p_0, q_0) lie on the normal curve, i.e. $F(p_0, q_0) = 0$, or

$$\frac{\alpha(\beta - 1)^2}{\gamma^2 - 4\alpha\beta} + \frac{\gamma[(\beta - 1)^2 (\alpha - 1)^2]^{1/2}}{\gamma^2 - 4\alpha\beta} + \frac{\beta(\alpha - 1)^2}{\gamma^2 - 4\alpha\beta} - (\alpha + 1) \left[\frac{(\beta - 1)^2}{\gamma^2 - 4\alpha\beta} \right]^{1/2}$$

$$- (\beta + 1) \left[\frac{(\alpha - 1)^2}{\gamma^2 - 4\alpha\beta} \right]^{1/2} + 1 = 0 \tag{2.4.8}$$

Thus an admissible value of $\gamma = \gamma_0(\alpha, \beta)$ must be a root of eq. (2.4.8), considered as a function of γ, and in addition fall within the proper γ ranges of cases n.5, n.8, and n.13 where $n = 1, 2, 3$ or 4. A graphical analysis of eq. (2.4.8) easily shows that this equation will have either zero or two positive roots, although neither root will necessarily be admissible.

For case 2 (see eq. (1.5.46) and Table 3) $0 < \alpha < 1$ and $\beta > 1$. Set $(\gamma^2 - 4\alpha\beta)F(p_0, q_0) = g(\gamma)$, then $g(\gamma)$ may be expressed in the form

$$g(\gamma) = (\beta - 1)(1 - \alpha)\{\gamma - (\alpha + \beta)\} + [\sqrt{\gamma^2 - 4\alpha\beta} - (\beta - \alpha)]^2. \tag{2.4.9}$$

Since $F(p_0, q_0) = 0$ for $\gamma = \gamma_0$, γ_0 will necessarily be a zero of $g(\gamma)$. By inspection $\gamma = (\alpha + \beta)$ is a zero of $g(\gamma)$, but this is not an admissible value of γ because $(\alpha + \beta) > (1 + \alpha\beta)$ for cases 2.5, 2.8 and 2.13. As previously indicated, $g(\gamma)$ will have two positive roots. Denote the second positive root of $g(\gamma)$ by $\gamma = \bar{\gamma}$. Now (by direct calculation) $g(\alpha + \beta) = 0$, $g'(\alpha + \beta) > 0$; hence $g(\bar{\gamma}) = 0$ while $g'(\bar{\gamma}) < 0$. It therefore follows that $\bar{\gamma} < (\alpha + \beta)$, use having been made of the fact that $g(+\infty) = +\infty$. Thus $2\sqrt{\alpha\beta} < \bar{\gamma} < (\alpha + \beta)$ (since by (2.4.6), $(\gamma^2 - 4\alpha\beta) > 0$) and $g(\gamma) > 0$ for $2\sqrt{\alpha\beta} < \gamma < \bar{\gamma}$. The question which must now be answered is whether or not $\gamma = \bar{\gamma}$ is an admissible γ value. By direct calculation $g(\alpha\beta + 1) > 0$, and since $\gamma = (\alpha\beta + 1)$ is the upper bound for the allowable γ range under cases 2.5, 2.8 and 2.13, it follows that $g(\gamma) > 0$ for these cases. Thus $\bar{\gamma}$ is not admissible since $\bar{\gamma} > (\alpha\beta + 1)$.

4. Bitangents Which Cross Both Coordinate Axes of the Normal Curve

For the γ range of interest, namely $2\sqrt{\alpha\beta} \leqslant \gamma \leqslant (1 + \alpha\beta)$, it has been shown that it is impossible for two inflection points (neither of which is on the p or q axes) to coalesce. In order to have inflection points, the outer branch of the normal curve must have real bitangents. Suppose, e.g., in cases 2.5 there is some value of γ, say $\gamma = \gamma_1$ for which a real bitangent exists. Now it is known that at the extreme values of γ, namely $\gamma = (\alpha + 1)$ and $\gamma = (1 + \alpha\beta)$ there is no bitangent with slope other than zero (for $\gamma = \alpha + 1$). Thus somewhere in the range $(\alpha + 1) < \gamma < (1 + \alpha\beta)$, the real bitangent must cease to exist, i.e. become a complex bitangent. Since there is continuous dependence of the outer branch of the normal curve on γ, such a point can only occur where two inflection points coalesce. But, as proven above, there is no such point for the γ range of case 2.5. The same can also be said of cases 2.8 and 2.13.

The above argument for the nonexistence of bitangents which cross both coordinate axes, can be repeated for cases 3.5, 3.8 and 3.13 of Table 4 where $\alpha > 1$ and $0 < \beta < 1$. This allows the following conclusion to be made: *the normal curve appropriate to the six cases 2.5, 2.8, 2.13, 3.5, 3.8 and 3.13 does not have (real) inflection points; hence the normal curve is always of class I.* The normal curve class will be defined below in section 6.

Finally the possibility of double inflection points under cases 1.5, 1.8, 1.13, 4.5, 4.8 and 4.13 must be investigated. Under case 1, α and β lie in the range $0 < \alpha < 1$, $0 < \beta < 1$. Again set

$$(\gamma^2 - 4\alpha\beta)F(p_0, q_0) = g(\gamma)$$

where $g(\gamma)$ may be placed in the form

$$g(\gamma) = (1 - \beta)(1 - \alpha)\{\gamma - (1 + \alpha\beta)\}\, [\sqrt{\gamma^2 - 4\alpha\beta} - (1 - \alpha\beta)]^2.$$

$$(2.4.10)$$

By inspection $\gamma = (1 + \alpha\beta)$ is a zero of $g(\gamma)$, but this is not an admissible γ value since it does not fall within the γ range appropriate to cases 1.5, 1.8 or 1.13. Since $F(p_0, q_0) = 0$, it follows that $g(\gamma) = 0$ and this latter equation becomes a quartic in γ, after the square root symbol is removed by squaring. Factoring the known root $\gamma = (1 + \alpha\beta)$ from the quartic, gives the cubic

$$f(\gamma) = \gamma^3 + l\gamma^2 + h\gamma + n = 0,$$

$$(2.4.11)$$

where

$$l = [(1 + \alpha\beta) + 2(1 - \beta)(1 - \alpha)] > 0, \qquad h = -(\alpha + \beta)[(1 - \alpha)$$
$$+ (1 - \beta) + 2\alpha\beta] < 0$$

and

$$n = -[(\alpha + \beta)^2(1 + \alpha\beta) + 16\alpha\beta(1 - \beta)(1 - \alpha)] < 0$$

From a consideration of the maximum and minimum points of $f(\gamma)$, it can be shown that $f(\gamma)$ has three real and distinct zeros, only one of which (say) $\gamma = \gamma_0(\alpha, \beta)$ is positive. Furthermore, a direct calculation reveals that γ_0 will lie in the appropriate γ

range for cases 1.5, 1.8 and 1.13. Thus $\gamma_0(\alpha, \beta)$ is an admissible γ value, so that if the α and β values of the triplet $(\alpha, \beta, \gamma_0)$ fall within cases 1.5, 1.8 or 1.13, the corresponding outer branch of the normal curve will have (four) double inflection points. An explicit expression for γ_0 is

$$\gamma_0(\alpha, \beta) = 2[(1 + \alpha\beta) - (\alpha + \beta)/3] \cos(\phi/3) - [(1 + \alpha\beta) - 2(\alpha + \beta)/3],$$

$$(2.4.12)$$

where

$$\cos\phi = \frac{-c}{2[(1 + \alpha\beta) - (\alpha + \beta)/3]^3} \quad \text{with} \quad 0 < \phi < \pi,$$

and $c = [2l^3 - 9lh + 27n]/27$. Note that $\gamma_0(\alpha, \beta) = \gamma_0(\beta, \alpha)$ and that

$$2\gamma_0(\alpha, \alpha) = -3(1 - \alpha)^2 + (1 + \alpha)\sqrt{9 - 14\alpha + 9\alpha^2}, \qquad (2.4.13)$$

which is the simplified equation for γ_0 under case 1.8.

For the γ range of interest under cases 1.5, 1.8 and 1.13, it has now been shown that it is possible for two inflection points (neither of which is on the p or q axes) to coalesce, provided $\gamma = \gamma_0(\alpha, \beta)$. In fact the p and q coordinates of the first quadrant coalesced inflection point are given in (2.4.6) and (2.4.7) with γ satisfying (2.4.12). In order to have inflection points, N_+ must have real bitangents. As γ approaches the value $(1 + \alpha\beta)$, the normal curve approaches the shape of two intersecting ellipses. Therefore for γ just below the value $(1 + \alpha\beta)$, N_+ will certainly have real bitangents. On the other hand as γ approaches the lower limit in cases 1.5, 1.8 and 1.13, no real bitangents (with slope other than zero or infinity) occur.

The analysis of cases 4.1, 4.5 and 4.13 closely parallels the above discussion, and in fact the same expression for $\gamma_0(\alpha, \beta)$ also applies here. The reference for this section is Payton [23]. See also Osipov [84] where a condition equivalent to (2.4.9) has been independently obtained. Some further simplifying approximations to γ_0 can be found in Musgrave and Payton [85].

5. Double Points of the Normal Curve

As mentioned in section 5 of Chapter 1, when the (α, β, γ) combination is such that

$$B^2(\theta) - 4A(\theta) = 0, \qquad (2.5.1)$$

for certain values of θ, then the Q operator ceases to be strictly hyperbolic. In these situations the normal curve will contain double points where N_+ and N_- come in contact with each other. There are six cases to be treated here, previously noted by (1.5.49)–(1.5.54).

Case 5) $0 < \alpha < 1,$ $0 < \beta < 1,$ $\gamma = (\alpha\beta + 1).$

N is composed of two ellipses: $\alpha p^2 + q^2 = 1$ and $p^2 + \beta q^2 = 1$ which intersect at the four double points of N given by

$$p = \pm \sqrt{\frac{1 - \beta}{1 - \alpha\beta}}, \qquad q = \pm \sqrt{\frac{1 - \alpha}{1 - \alpha\beta}}. \qquad (2.5.2)$$

Case 6) $\alpha > 1$, $\beta > 1$, $\gamma = (\alpha\beta + 1)$

Same remarks as Case 5).

Case 7) $\alpha = 1$, $\beta \neq 1$, $-2\sqrt{\beta} < \gamma \leqslant (1 + \beta)$.

There will be two double points on N at

$$p = \pm 1, \qquad q = 0. \tag{2.5.3}$$

From (2.3.2) the slope of the normal curve is given by

$$\frac{dq}{dp} = -\frac{p[2p^2 + \gamma q^2 - 2]}{q[2\beta q^2 + \gamma p^2 - (\beta + 1)]} \tag{2.5.4}$$

This expression becomes indeterminate at $p = \pm 1$, $q = 0$ where N_+ and N_- have corners. Evaluation of this indeterminate form gives

$$\left(\frac{dq}{dp}\right)^2 = \frac{4}{\beta + 1 - \gamma}, \tag{2.5.5}$$

and it is easily shown that

$$\text{on } N_-: \qquad \lim_{\theta \to 2\pi - 0} \frac{dq}{dp} = \frac{2}{\sqrt{\beta + 1 - \gamma}}, \qquad \lim_{\theta \to 0+0} \frac{dq}{dp} = -\frac{2}{\sqrt{\beta + 1 - \gamma}},$$

$$\tag{2.5.6}$$

$$\text{on } N_+: \qquad \lim_{\theta \to 2\pi - 0} \frac{dq}{dp} = -\frac{2}{\sqrt{\beta + 1 - \gamma}}, \qquad \lim_{\theta \to 0+0} \frac{dq}{dp} = \frac{2}{\sqrt{\beta + 1 - \gamma}}.$$

$$\tag{2.5.7}$$

Thus N_+ has a local 'minimum' (henceforth referred to as 'MIN') at $\theta = 0$ (and $\theta = \pi$ by symmetry). At $\theta = \pi/2$, $N_\pm$ are smooth so that the local properties of the two branches of N may be deduced from (2.3.5). N_-will have a corner maximum at $\theta = 0$ and a smooth maximum at $\theta = \pi/2$. Furthermore the analysis of the previous section (in particular eqn. (2.4.7) with $\alpha = 1$) indicates that under Case 7) there can be no finite (nonzero) slope bitangents to N_+. Thus N_+ can not have a local minimum for some θ-value in the range $0 < \theta < \pi/2$ when $\alpha = 1$.

Case 8) $\alpha \neq 1$, $\beta = 1$, $-2\sqrt{\alpha} < \gamma \leqslant (1 + \alpha)$.

There will be two double points on N at

$$p = 0, \qquad q = \pm 1. \tag{2.5.8}$$

The discussion follows closely that of Case 7) above. In particular N_+ has a local 'minimum' at $\theta = \pi/2$ and $\theta = 3\pi/2$.

Case 9) $\alpha = \beta = 1$, $-2 < \gamma < 2$.

N is composed of two ellipses: $\mu_1^2 p'^2 + \mu_2^2 q'^2 = 1$ and $\mu_2^2 p'^2 + \mu_1^2 q'^2 = 1$ where $p = (p' - q')/\sqrt{2}$, $q = (p' + q')/\sqrt{2}$, $\mu_1^2 = (2 - \sqrt{2 - \gamma})/2$ and $\mu_2^2 = (2 + \sqrt{2 - \gamma})/2$.

These will be four double points on N at

$$p = \pm 1, \qquad q = 0 \quad \text{and} \quad p = 0, \qquad q = \pm 1. \tag{2.5.9}$$

Case 10) $\alpha = \beta = 1, \qquad \gamma = 2.$

This is a degenerate case in which all points of N are double points since N is composed of two coinciding circles

$$N: (p^2 + q^2 - 1)^2 = 0. \tag{2.5.10}$$

Under Cases 5), 6) and 10) of this section, κ (as defined in (1.6.4)) is zero. For these cases the two dimensional equations of motion (as given in section 1.6) uncouple so that the displacement components are governed by a second order equation. The same remarks apply to Case 7) when $\gamma = 1 + \beta$ and Case 8) when $\gamma = 1 + \alpha$.

The discussion in this section follows that of Payton [12].

6. Classification of the Normal Curve Shape

After the preceding lengthy discussion of the geometry of the normal curve, this curve will now be classified according to its shape. The distinguishing features are the number and location of inflection points on N_+. For α, β and γ values for which N is free of double points, there are five classes.

Class I N_+ has no bitangents.
Class II N_+ has two vertical (but no horizontal) bitangents.
Class III N_+ has two horizontal (but no vertical) bitangents.
Class IV N_+ has two horizontal and two vertical bitangents.
Class V N_+ has four bitangents, none of which are horizontal or vertical.

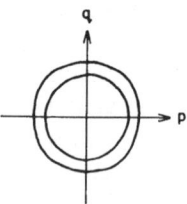

Figure 1.   Class I normal curve. ($\alpha = 0.64, \beta = 0.64, \gamma = 1.2$)

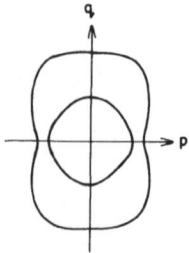

Figure 2.   Class II normal curve. ($\alpha = 0.64, \beta = 0.25, \gamma = 0.42$)

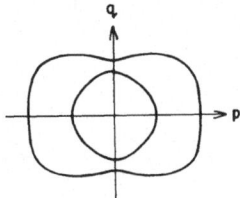

Figure 3.   Class III normal curve. ($\alpha = 0.25, \beta = 0.64, \gamma = 0.42$)

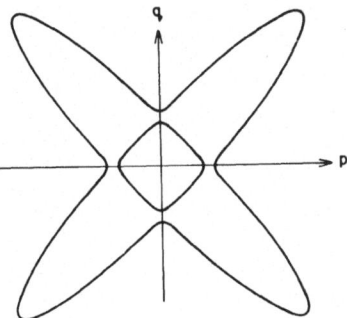

Figure 4.   Class IV normal curve. ($\alpha = 0.64, \beta = 0.64, \gamma = -1$)

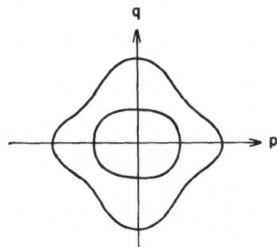

Figure 5.   Class V normal curve. ($\alpha = 4, \beta = 6, \gamma = 20$)

Typical curves of N for these classes are shown in figures 1–5. Tables 2–5 classify the normal curve for Cases 1)–4) of section 1.5.

When α, β and γ fall under Cases 5)–10), the normal curve will contain double points. Except for Cases 7) and 8) N is composed of intersecting ellipses (or circles as in Case 10)) and requires no figures, since the description in the previous section should be sufficient. Typical curves for N for Cases 7.1, 7.2 and 7.3 are shown in figures 6–8. The curves for Cases 7.5–7.7 exhibit the same features. The normal curves for Case 8) can be obtained by rotating figures 6–8 by 90°. Table 6 classifies N for Cases 7) and 8).

Table 2.   Wave Front and Normal Curve Classification for Case 1 $[0 < \alpha < 1,\ 0 < \beta < 1,$
$-2\sqrt{\alpha\beta} < \gamma < (\alpha\beta + 1)]$

| | | | Local Behaviour on N_+ | | |
		γ-range	$\theta = 0$	$\theta = \pi/2$	Class
	1.1	$-2\sqrt{\alpha\beta} < \gamma < \beta(\alpha + 1)$	MIN	MIN	IV
	1.2	$\gamma = \beta(\alpha + 1)$	MIN	CINFL	II
	1.3	$\beta(\alpha + 1) < \gamma < \alpha(\beta + 1)$	MIN	MAX	II
$\alpha > \beta$	1.4	$\gamma = \alpha(\beta + 1)$	CINFL	MAX	I
	1.5.1	$\alpha(\beta + 1) < \gamma < \gamma_0(\alpha, \beta)$	MAX	MAX	I
	1.5.2	$\gamma = \gamma_0(\alpha, \beta)$	MAX	MAX	I
	1.5.3	$\gamma_0(\alpha, \beta) < \gamma < (\alpha\beta + 1)$	MAX	MAX	V
	1.6	$-2\alpha < \gamma < \alpha(\alpha + 1)$	MIN	MIN	IV
	1.7	$\gamma = \alpha(\alpha + 1)$	CINFL	CINFL	I
$\alpha = \beta$	1.8.1	$\alpha(\alpha + 1) < \gamma < \gamma_0(\alpha, \alpha)$	MAX	MAX	I
	1.8.2	$\gamma = \gamma_0(\alpha, \alpha)$	MAX	MAX	I
	1.8.3	$\gamma_0(\alpha, \alpha) < \gamma < (\alpha^2 + 1)$	MAX	MAX	V
	1.9	$-2\sqrt{\alpha\beta} < \gamma < \alpha(\beta + 1)$	MIN	MIN	IV
	1.10	$\gamma = \alpha(\beta + 1)$	CINFL	MIN	III
	1.11	$\alpha(\beta + 1) < \gamma < \beta(\alpha + 1)$	MAX	MIN	III
$\alpha < \beta$	1.12	$\gamma = \beta(\alpha + 1)$	MAX	CINFL	I
	1.13.1	$\beta(\alpha + 1) < \gamma < \gamma_0(\alpha, \beta)$	MAX	MAX	I
	1.13.2	$\gamma = \gamma_0(\alpha, \beta)$	MAX	MAX	I
	1.13.3	$\gamma_0(\alpha, \beta) < \gamma < (\alpha\beta + 1)$	MAX	MAX	V

Table 3.   Wave Front and Normal Curve Classification for Case 2, $[0 < \alpha < 1,\ \beta > 1,$
$-2\sqrt{\alpha\beta} < \gamma \leqslant (\alpha\beta + 1)]$

| | | | Local Behaviour on N_+ | | |
		γ-range	$\theta = 0$	$\theta = \pi/2$	Class
	2.1	$-2\sqrt{\alpha\beta} < \gamma < \alpha(\beta + 1)$	MIN	MIN	IV
	2.2	$\gamma = \alpha(\beta + 1)$	CINFL	MIN	III
$\alpha\beta < 1$	2.3	$\alpha(\beta + 1) < \gamma < (\alpha + 1)$	MAX	MIN	III
	2.4	$\gamma = (\alpha + 1)$	MAX	CINFL	I
	2.5	$(\alpha + 1) < \gamma \leqslant (1 + \alpha\beta)$	MAX	MAX	I
	2.6	$-2 < \gamma < (1 + \alpha)$	MIN	MIN	IV
$\alpha\beta = 1$	2.7	$\gamma = (1 + \alpha)$	CINFL	CINFL	I
	2.8	$(1 + \alpha) < \gamma \leqslant 2$	MAX	MAX	I
	2.9	$-2\sqrt{\alpha\beta} < \gamma < (\alpha + 1)$	MIN	MIN	IV
	2.10	$\gamma = (\alpha + 1)$	MIN	CINFL	II
$\alpha\beta > 1$	2.11	$(\alpha + 1) < \gamma < \alpha(\beta + 1)$	MIN	MAX	II
	2.12	$\gamma = \alpha(\beta + 1)$	CINFL	MAX	I
	2.13	$\alpha(\beta + 1) < \gamma \leqslant (1 + \alpha\beta)$	MAX	MAX	I

Table 4. Wave Front and Normal Curve Classification for Case 3, $[\alpha > 1, \ 0 < \beta < 1,$ $-2\sqrt{\alpha\beta} < \gamma \leqslant (1 + \alpha\beta)]$

		γ-range	Local Behavior on N_+		
			$\theta = 0$	$\theta = \pi/2$	Class
	3.1	$-2\sqrt{\alpha\beta} < \gamma < \beta(\alpha + 1)$	MIN	MIN	IV
	3.2	$\gamma = \beta(\alpha + 1)$	MIN	CINFL	II
$\alpha\beta < 1$	3.3	$\beta(\alpha + 1) < \gamma < (\beta + 1)$	MIN	MAX	II
	3.4	$\gamma = (\beta + 1)$	CINFL	MAX	I
	3.5	$(\beta + 1) < \gamma \leqslant (1 + \alpha\beta)$	MAX	MAX	I
	3.6	$-2 < \gamma < (\beta + 1)$	MIN	MIN	IV
$\alpha\beta = 1$	3.7	$\gamma = (\beta + 1)$	CINFL	CINFL	I
	3.8	$(\beta + 1) < \gamma \leqslant 2$	MAX	MAX	I
	3.9	$-2\sqrt{\alpha\beta} < \gamma < (\beta + 1)$	MIN	MIN	IV
$\alpha\beta > 1$	3.10	$\gamma = (\beta + 1)$	CINFL	MIN	III
	3.11	$(\beta + 1) < \gamma < \beta(\alpha + 1)$	MAX	MIN	III
	3.12	$\gamma = \beta(\alpha + 1)$	MAX	CINFL	I
	3.13	$\beta(\alpha + 1) < \gamma \leqslant (1 + \alpha\beta)$	MAX	MAX	I

Table 5. Wave Front and Normal Curve Classification for Case 4. $[\alpha > 1, \ \beta > 1,$ $-2\sqrt{\alpha\beta} < \gamma < (\alpha\beta + 1)]$

		γ-Range	Local Behavior on N_+		
			$\theta = 0$	$\theta = \pi/2$	Class
	4.1	$-2\sqrt{\alpha\beta} < \gamma < (\beta + 1)$	MIN	MIN	IV
	4.2	$\gamma = (\beta + 1)$	CINFL	MIN	III
	4.3	$(\beta + 1) < \gamma < (\alpha + 1)$	MAX	MIN	III
$\alpha > \beta$	4.4	$\gamma = (\alpha + 1)$	MAX	CINFL	I
	4.5.1	$(\alpha + 1) < \gamma < \gamma_0(\alpha, \beta)$	MAX	MAX	I
	4.5.2	$\gamma = \gamma_0(\alpha, \beta)$	MAX	MAX	I
	4.5.3	$\gamma_0(\alpha, \beta) < \gamma < (\alpha\beta + 1)$	MAX	MAX	V
	4.6	$-2\alpha < \gamma < (\alpha + 1)$	MIN	MIN	IV
	4.7	$\gamma = (\alpha + 1)$	CINFL	CINFL	I
$\alpha = \beta$	4.8.1	$(\alpha + 1) < \gamma < \gamma_0(\alpha, \alpha)$	MAX	MAX	I
	4.8.2	$\gamma = \gamma_0(\alpha, \alpha)$	MAX	MAX	I
	4.8.3	$\gamma_0(\alpha, \alpha) < \gamma < (\alpha^2 + 1)$	MAX	MAX	V
	4.9	$-2\sqrt{\alpha\beta} < \gamma < (\alpha + 1)$	MIN	MIN	IV
	4.10	$\gamma = (\alpha + 1)$	MIN	CINFL	II
$\alpha < \beta$	4.11	$(\alpha + 1) < \gamma < (\beta + 1)$	MIN	MAX	II
	4.12	$\gamma = (\beta + 1)$	CINFL	MAX	I
	4.13.1	$(\beta + 1) < \gamma < \gamma_0(\alpha, \beta)$	MAX	MAX	I
	4.13.2	$\gamma = \gamma_0(\alpha, \beta)$	MAX	MAX	I
	4.13.3	$\gamma_0(\alpha, \beta) < \gamma < (\alpha\beta + 1)$	MAX	MAX	V

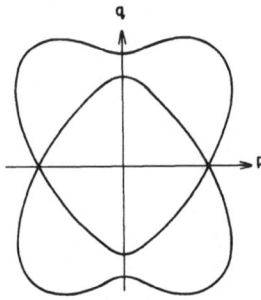

Figure 6.   Normal curve for case 7.1. ($\alpha = 1, \beta = 0.64, \gamma = 0.42$)

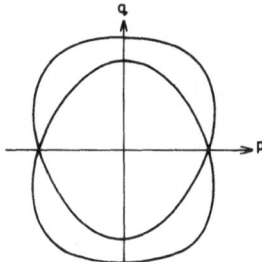

Figure 7.   Normal curve for case 7.2. ($\alpha = 1, \beta = 0.64, \gamma = 1.28$)

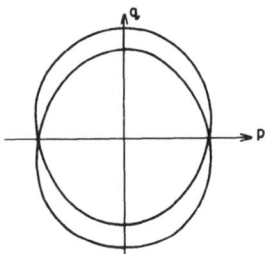

Figure 8.   Normal curve for case 7.3. ($\alpha = 1, \beta = 0.64, \gamma = 1.50$)

Table 6.   Wave Front and Normal Curve Classification for Cases 7 and 8.

| | | | Local Behavior on N_+ | |
		γ-Range	$\theta = 0$	$\theta = \pi/2$
$\alpha = 1$	7.1	$-2\sqrt{\beta} < \gamma < 2\beta$	'MIN'	MIN
$0 < \beta < 1$	7.2	$\gamma = 2\beta$	'MIN'	CINFL
	7.3	$2\beta < \gamma < (1 + \beta)$	'MIN'	MAX
$\alpha = 1$	7.5	$-2\sqrt{\beta} < \gamma < 2$	'MIN'	MIN
$\beta > 1$	7.6	$\gamma = 2$	'MIN'	CINFL
	7.7	$2 < \gamma < (1 + \beta)$	'MIN'	MAX
$\beta = 1$	8.1	$-2\sqrt{\alpha} < \gamma < 2\alpha$	MIN	'MIN'
$0 < \alpha < 1$	8.2	$\gamma = 2\alpha$	CINFL	'MIN'
	8.3	$2\alpha < \gamma < (1 + \alpha)$	MAX	'MIN'
$\beta = 1$	8.5	$-2\sqrt{\alpha} < \gamma < 2$	MIN	'MIN'
$\alpha > 1$	8.6	$\gamma = 2$	CINFL	'MIN'
	8.7	$2 < \gamma < (1 + \alpha)$	MAX	'MIN'

7. Solution of the Characteristic Partial Differential Equation

In this section the solution to the nonlinear, first order partial differential equation

$$F(p,q) = 0, \tag{2.7.1}$$

where

$$p = \psi_z \quad \text{and} \quad q = \psi_y, \tag{2.7.2}$$

will be obtained by the method of *characteristic strips*. This is the standard method for solving an equation of the type (2.7.1) and the theory of this method is well covered in volume 2 of Courant and Hilbert [18]. The solution method used on a similar problem concerning wave fronts for electromagnetic waves in anisotropic media by Kline and Kay [19] is somewhat more relevant and will be followed here. The following discussion, although brief, is complete. In the next section an alternate (and more physical) interpretation is given to the solution of (2.7.1).

Recall from section (2.2) that the wave front curve W has the equation

$$\psi(y,z) - \tau = 0 \tag{2.7.3}$$

provided ψ satisfies (2.7.1) and the initial condition

$$\psi(0,0) = 0. \tag{2.7.4}$$

The latter condition being appropriate to a wave front which emanates at time $\tau = 0$ from a source which is localized at the origin. By the method of characteristic strips the variables z, y, p, q and ψ are treated on an equal footing along a curve $z(s), y(s), \psi(s)$ in 3-space (z, y, ψ). This curve is called a characteristic curve (or more properly a bicharacteristic curve) and since it carries with it at each point a normal vector $(p(s), q(s), -1)$, the combination generates an infinitesimal strip or characteristic strip. The totality of all such strips form a surface whose trace in the plane $\psi = \tau$ gives the wave front curve at the time τ.

The ordinary differential equations associated with the strip are

$$\frac{dz}{ds} = F_p, \qquad \frac{dy}{ds} = F_q, \qquad \frac{dp}{ds} = -(F_\psi p + F_z) = 0,$$

$$\frac{dq}{ds} = -(F_\psi q + F_y) = 0$$

and

$$\frac{d\psi}{ds} = pF_p + qF_q, \tag{2.7.5}$$

with initial conditions appropriate to (2.7.4)

$$z(0) = 0, \qquad y(0) = 0, \qquad p(0) = p_0, \qquad q(0) = q_0 \quad \text{and}$$

$$\psi(0) = 0. \tag{2.7.6}$$

Since $dp/ds = 0$ and $dq/ds = 0$, it follows from (2.7.6) that

$$p = p_0 = \text{constant} \quad \text{and} \quad q = q_0 = \text{constant} \tag{2.7.7}$$

so that p and q are constant along a characteristic curve, hence upon further integration of (2.7.5)

$$z = F_p s, \qquad y = F_q s \quad \text{and} \quad \psi = (pF_p + qF_q)s, \tag{2.7.8}$$

or

$$\psi = pz + qy. \tag{2.7.9}$$

From (2.7.8) it follows (since the solution depends linearly on s) that the characteristics are straight lines through the origin (in 3-space). In (2.7.9) the form for ψ is deceptive since p and q are not independent but are in fact related through (2.7.1). Equation (2.7.9) thus gives an implicit solution to the problem. It can, in theory, be made explicit as follows. From (2.7.8)

$$s = z/F_p = y/F_q.$$

Hence, invert the two equations

$$z/F_p = y/F_q \quad \text{and} \quad F(p,q) = 0, \tag{2.7.10}$$

to determine

$$p = f(z,y) \quad \text{and} \quad q = g(z,y). \tag{2.7.11}$$

Then from (2.7.9) and (2.7.3), the equation for the wave front is

$$zf(z,y) + yg(z,y) = \tau. \tag{2.7.12}$$

But all this is easier said than done.

An alternate solution form for ψ will now be established. It will be much easier to use and leads directly to the solution construction of the next section. From (2.7.8) and (2.7.3)

$$\psi = (pF_p + qF_q)s = \tau \quad \text{so that} \quad s = \frac{\tau}{pF_p + qF_q}. \tag{2.7.13}$$

(Note that this method of solution breaks down if $F_p^2 + F_q^2 = 0$, as it does at a corner of N). But from (2.7.8) $z = F_p s$ so that

$$z = \frac{\tau F_p}{pF_p + qF_q} \quad \text{and similarly} \quad y = \frac{\tau F_q}{pF_p + qF_q} \tag{2.7.14}$$

If the right side of these equations is expressed in terms of θ by

$$p = R(\theta)\cos\theta, \qquad q = R(\theta)\sin\theta, \tag{2.7.15}$$

from (2.2.8) where $R(\theta)$ is given in (2.2.6), then a θ parametric representation of W is obtained. Toward this end, from (2.7.15)

$$dp = (R'\cos\theta - R\sin\theta)d\theta, \qquad dq = (R'\sin\theta + R\cos\theta)d\theta,$$

so

$$\frac{dq}{dp} = \frac{R' \sin\theta + R\cos\theta}{R'\cos\theta - R\sin\theta} = \frac{\dfrac{R'}{R^2}\sin\theta + \dfrac{\cos\theta}{R}}{\dfrac{R'}{R^2}\cos\theta - \dfrac{\sin\theta}{R}}$$

Or

$$\frac{dq}{dp} = \frac{-V'\sin\theta + V\cos\theta}{-V'\cos\theta - V\sin\theta} \cdot \qquad (2.7.16)$$

But from (2.7.1)

$$F_p dp + F_q dq = 0 \quad \text{or} \quad \frac{dq}{dp} = -\frac{F_p}{F_q} \qquad (2.7.17)$$

Hence, from (2.7.14)

$$z = \frac{\tau F_p}{pF_p + qF_q} = \frac{\tau}{p + q\dfrac{F_q}{F_p}} = \frac{\tau}{p - \dfrac{q}{dq/dp}} =$$

$$\frac{\tau}{\dfrac{\cos\theta}{V} - \dfrac{\sin\theta/V}{\dfrac{-V'\sin\theta + V\cos\theta}{-V'\cos\theta - V\sin\theta}}} \qquad (2.7.18)$$

where

$$V(\theta) = 1/R(\theta). \qquad (2.7.19)$$

The last expression in (2.7.18) simplifies to

$$z = \tau(V\cos\theta - V'\sin\theta), \qquad (2.7.20)$$

and similarly for y,

$$y = \tau(V\sin\theta + V'\cos\theta) \qquad (2.7.21)$$

Equations (2.7.20) and (2.7.21) constitute a parametric solution for the wave front curve. As θ ranges over the interval $0 \leqslant \theta < 2\pi$, the point (z, y) (for a given τ value) will trace out a curve W. W will have two branches W_+ for $V = V_+ = 1/R_+$ and W_- for $V = V_- = 1/R_-$. Thus corresponding to the two branches of N, there are two branches of W. Because of the inversion (2.7.19), the branch W_+ of W will lie within the branch W_- of W. Further geometric features of W will be discussed in the next section.

8. Wave Front Construction as Envelope of Line Waves

As shown in (2.7.9), ψ satisfies

$$\psi = pz + qy = \tau, \tag{2.8.1}$$

provided p and q are related by $F(p, q) = 0$. Representing p and q in polar coordinates gives

$$zR \cos \theta + yR \sin \theta = \tau,$$

or

$$z \cos \theta + y \sin \theta = V(\theta)\tau, \tag{2.8.2}$$

where $V(\theta) = 1/R(\theta)$. Equation (2.8.2) represents a set of straight lines (in (z, y) space) which constitute a one parameter (θ) family of solutions to the characteristic equation (2.2.2), provided the function $R(\theta)$ is given by (2.2.6). At a given positive time τ, eqn. (2.8.2) describes a line in (z, y) space, the normal to which has components $\cos \theta$ and $\sin \theta$ in the z and y directions respectively. Thus the normal vector to this line is a unit vector in the θ-direction. Furthermore, the line is located a distance $V(\theta)\tau$, measured along the normal vector, from the origin. Hence the line moves in time, with speed $V(\theta)$, normal to itself away from the origin. For this reason (2.8.2) is referred to as a *line wave*. The above interpretation is valid for any θ-direction, however it should be noted that the wave speed will differ as θ varies due to the anisotropic nature of the medium.

According to Huyghens' construction (see Courant and Hilbert [18]) the general form of the wave front emitted by a concentrated disturbance at the origin at time $\tau = 0$, is given by the envelope of the one parameter family of line waves (2.8.2). The established procedure for finding the envelope of the family of straight lines (2.8.2) is the elimination of θ between this equation and its derivative

$$- z \sin \theta + y \cos \theta = V'(\theta)\tau. \tag{2.8.3}$$

Alternatively equations (2.8.2) and (2.8.3) may be solved for

$$\bar{z} = z/\tau \quad \text{and} \quad \bar{y} = y/\tau, \tag{2.8.4}$$

(which, by virtue of (1.5.23) are dimensionless variables) as functions of θ, thus giving the parametric representation of the wave front

$$\bar{z}(\theta) = V(\theta) \cos \theta - V'(\theta) \sin \theta,$$

and

$$\bar{y}(\theta) = V(\theta) \sin \theta + V'(\theta) \cos \theta. \tag{2.8.5}$$

Equations (2.8.5) agree with (2.7.20) and (2.7.21), thus justifying the construction of Huyghens.

In summary, the wave front curve W will consist of two branches given parametrically by

$$W_+ : \bar{z}(\theta) = \bar{z}_+(\theta), \qquad \bar{y}(\theta) = \bar{y}_+(\theta), \qquad 0 \leqslant \theta < 2\pi \tag{2.8.6}$$

and

$$W_- : \bar{z}(\theta) = \bar{z}_-(\theta), \qquad \bar{y}(\theta) = \bar{y}_-(\theta), \qquad 0 \leqslant \theta < 2\pi. \tag{2.8.7}$$

Upon performing the differentiation indicated in (2.8.5), explicit expressions for $\bar{z}_\pm$ and $\bar{y}_\pm$ are

$$\bar{z}_\pm(\theta) = \frac{\cos\theta}{2AR_\pm}\left[2\alpha \cos^2\theta + \gamma \sin^2\theta \pm \frac{\sin^2\theta (k_1 \cos^2\theta - k_2 \sin^2\theta)}{\sqrt{B^2 - 4A}} \right],$$

$$\tag{2.8.8}$$

and

$$\bar{y}_\pm(\theta) = \frac{\sin\theta}{2AR_\pm}\left[2\beta \sin^2\theta + \gamma \cos^2\theta \mp \frac{\cos^2\theta (k_1 \cos^2\theta - k_2 \sin^2\theta)}{\sqrt{B^2 - 4A}} \right]$$

$$\tag{2.8.9}$$

where $k_1 = 2\alpha(\beta + 1) - \gamma(\alpha + 1)$ and $k_2 = 2\beta(\alpha + 1) - \gamma(\beta + 1)$. The expressions $A(\theta)$ and $B(\theta)$ were defined in eqns. (1.5.26) and (1.5.27).

Duff [24], has noted another direct method for constructing W as follows: 'let π be a tangent line to N at P, and Q the foot of the normal from the origin to π. Then let R be the point inverse to Q with respect to the unit circle; R lies on W.' Duff goes on to state that 'reciprocally, the same construction yields P on N given R on W.' A proof of this construction will now be given.

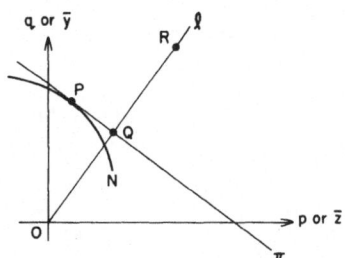

Figure 9.   Construction of the wave curve from the normal curve.

The point P on N has coordinates

$$P: \left(\frac{\cos\theta}{V}, \frac{\sin\theta}{V} \right) = (p_0, q_0). \tag{2.8.10}$$

The equation of the line tangent to N at P is

$$\pi : (q - q_0) = m(p - p_0) \tag{2.8.11}$$

where the slope m is given by (2.7.16). The equation of the normal line to π, which passes through the origin is

35

$$l: q = -\frac{1}{m} p. \tag{2.8.12}$$

At Q, l and π intersect so that

$$-\frac{1}{m} p - q_0 = mp - mp_0 \quad \text{or} \quad p = \frac{mp_0 - q_0}{m + 1/m}. \tag{2.8.13}$$

Hence, from (2.8.12) the point Q has the position vector

$$\mathbf{OQ} = \left(\frac{mp_0 - q_0}{m + 1/m}, -\frac{mp_0 - q_0}{m^2 + 1}\right) = \frac{mp_0 - q_0}{m^2 + 1} (m, -1). \tag{2.8.14}$$

So

$$|\mathbf{OQ}| = -\frac{mp_0 - q_0}{\sqrt{m^2 + 1}} \quad \text{assuming} \quad (mp_0 - q_0) < 0. \tag{2.8.15}$$

Since $|\mathbf{OQ}| \cdot |\mathbf{OR}| = 1$, it follows that R has coordinates

$$R: \frac{-1}{|\overline{\mathbf{OQ}}|} \frac{(m, -1)}{\sqrt{m^2 + 1}} = \frac{(m, -1)}{mp_0 - q_0}. \tag{2.8.16}$$

But,

$$mp_0 - q_0 = \frac{1}{-V' \cos \theta - V \sin \theta}, \tag{2.8.17}$$

hence

$$R: (V \cos \theta - V' \sin \theta, V \sin \theta + V' \cos \theta) \tag{2.8.18}$$

which agrees with (2.8.5).

Two important geometrical features of W_+ will now be proved.

(i) A bitangent on N_+ corresponds to a double point on W_+.

$$\text{At } (p_1, q_1): \; p - \frac{q}{dq/dp} = p_1 - \frac{q_1}{m} = p_1 - \frac{1}{m}(mp_1 + b) = -\frac{b}{m}. \tag{2.8.19}$$

$$\text{At } (p_2, q_2): \; p - \frac{q}{dq/dp} = p_2 - \frac{q_2}{m} = p_2 - \frac{1}{m}(mp_2 + b) = -\frac{b}{m}. \tag{2.8.20}$$

But on W_+, from (2.7.14)

$$\bar{z}_+ = \frac{F_p}{pF_p + qF_q} = \frac{1}{p - \dfrac{q}{dq/dp}} = \frac{1}{-b/m} \quad \text{at} \quad (p_1, q_1) \quad \text{and} \quad (p_2, q_2). \tag{2.8.21}$$

Also at (p_1, q_1),

$$q - p\frac{dq}{dp} = q_1 - mp_1 = b,\qquad(2.8.22)$$

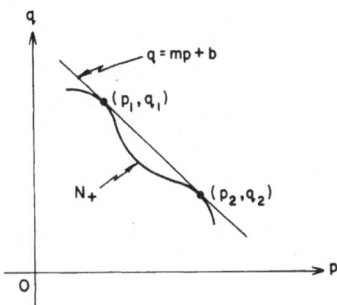

Figure 10. Normal curve showing a bitangent line.

and at (p_2, q_2)

$$q - p\frac{dq}{dp} = q_2 - mp_2 = b.\qquad(2.8.23)$$

Again, on W_+, from (2.7.14)

$$\bar{y}_+ = \frac{F_q}{pF_p + qF_q} = \frac{1}{q - p\dfrac{dq}{dp}} = \frac{1}{b}\quad\text{at}\quad(p_1, q_1)\quad\text{and}\quad(p_2, q_2).$$

$$(2.8.24)$$

Thus the two distinct points (p_1, q_1) and (p_2, q_2) on N_+ map into the same point $(-m/b, 1/b)$ on W_+.

(ii) An inflection point on N_+ corresponds to a critical point (cusp) on W_+.

The curvature of a plane curve given parametrically by

$$p = \frac{\cos\theta}{V(\theta)}\quad\text{and}\quad q = \frac{\sin\theta}{V(\theta)}$$

is

$$c = \frac{q''p' - p''q'}{[p'^2 + q'^2]^{3/2}}\qquad(2.8.25)$$

Performing the indicated differentiation and simplifying gives

$$c = \frac{V^3(V + V'')}{(V^2 + V'^2)^{3/2}}\qquad(2.8.26)$$

Recall that an inflection point of a plane curve is a point where the curvature changes sign. Hence at a point where $(V_+ + V_+'')$ changes sign N_+ will have an inflection point. Since N_- is convex, it contains no inflection points. From (2.8.5) by differentiation

$$\bar{z}_+' = -(V_+ + V_+'')\sin\theta,\qquad(2.8.27)$$

37

and

$$\bar{y}'_+ = (V_+ + V''_+) \cos \theta, \tag{2.8.28}$$

Now at a critical point of W_+, $z'_+ = 0$ and $\bar{y}'_+ = 0$, which from (2.8.27) and (2.8.28) implies that $(V_+ + V''_+) = 0$. From (2.8.26) this proves the correspondence between an inflection point on N_+ and a critical point on W_+. Incidentally this correspondence and the striking geometrical feature it gives rise to on W_+ is the reason so much effort was devoted to establishing a criterion for the existence of inflection points on N_+ in sections (2.3) and (2.4).

In connection with the material of this section, much good work has been done by Musgrave [20] and Duff [21].

9. Classification of the Wave Front Curve when Q is Strictly Hyperbolic

Following the normal curve classification of section 2.6, the wave front curve can now be easily classified according to the location of the cusps on W_+ associated with the inflection points on N_+. Again there are five classes.

Class I W_+ has no cuspidal triangles.
Class II W_+ has two (and only two) cuspidal triangles which are centered on the $\bar{z}$ axis.
Class III W_+ has two (and only two) cuspidal triangles which are centered on the $\bar{y}$ axis.
Class IV W_+ has four cuspidal triangles, two centered on the $\bar{z}$ axis and two centered on the $\bar{y}$ axis.
Class V W_+ has four cuspidal triangles, none of which cross the coordinate axes.

Typical curves of W for the above classes are shown in figures 11–15. Here the same values of α, β and γ are used as were used in figures 1–5 respectively. The single

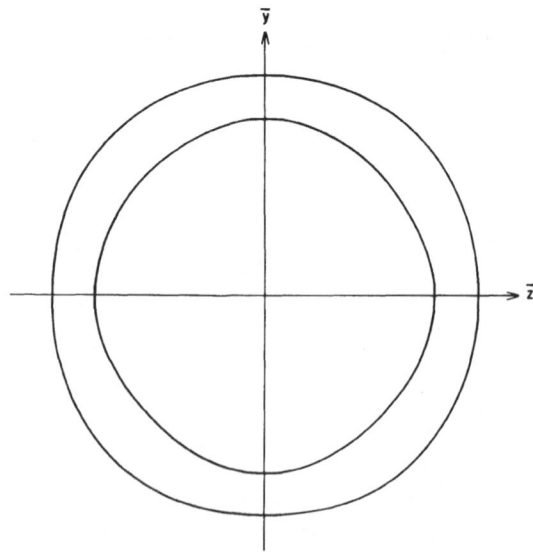

Figure 11. Class I wave front curve. ($\alpha = 0.64, \beta = 0.64, \gamma = 1.2$)

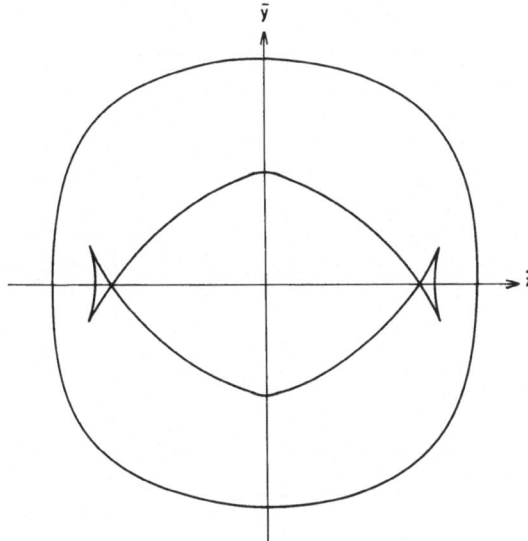

Figure 12. Class II wave front curve. ($\alpha = 0.64, \beta = 0.25, \gamma = 0.42$)

most important geometric feature of W is the existence (or nonexistence) of cusps on W_+. This in turn stems from the existence (or nonexistence) of inflection points on N_+. This remarkable morphogenesis of the wave front caused by a relative variation in the (α, β, γ) combination is due to the anisotropic nature of the elastic solid and has no counterpart in the case of isotropic elasticity. The multiple covering of the (z, y) plane by the base bicharacteristics of W (induced by the fold in the wave front cone in the 3-space (z, y, τ) has lead Thom [25] to refer to such a phenomenon as a cusp catastrophe.

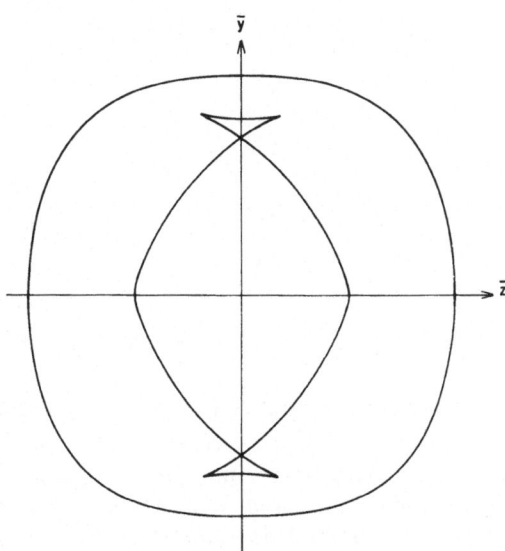

Figure 13. Class III wave front curve. ($\alpha = 0.25, \beta = 0.64, \gamma = 0.42$)

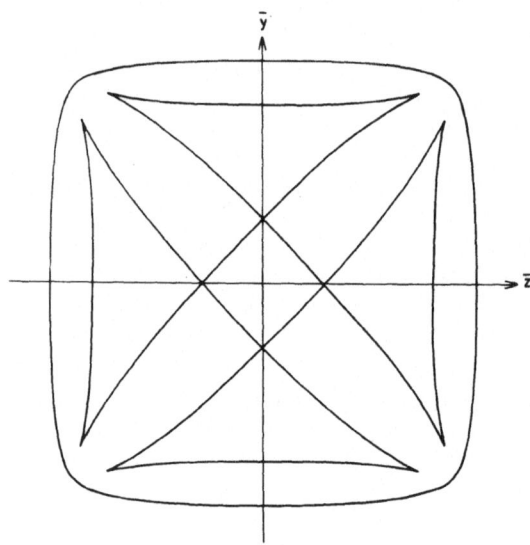

Figure 14.   Class IV wave front curve. ($\alpha = 0.64, \beta = 0.64, \gamma = -1$)

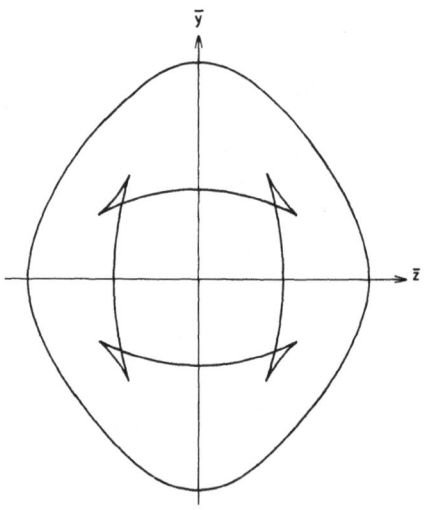

Figure 15.   Class V wave front curve. ($\alpha = 4, \beta = 6, \gamma = 20$)

10. Normal and Wave Front Curves for some Hexagonal Materials

Table 7 gives the nondimensional crystal constants α, β, γ and δ for the hexagonal crystals of Table 1. Note that for all these crystals $\alpha > 1$ and $\beta > 1$. Also listed in Table 7 are the values of $\gamma_0(\alpha, \beta)$ where appropriate. $\gamma_0(\alpha, \beta)$ was calculated from eqn. (2.4.12). Finally the case number (from Table 5) and class of N and W are given.

Equations (2.2.8) for $N_\pm$ and (2.8.8)–(2.8.9) for $W_\pm$ were programmed on a TEKTRONIX 4051 computer (by M. Bouscher) for display on a TEKTRONIX 4610

Table 7.   Normal and Wave Front Curve Classification For Some Hexagonal Crystals.

Crystal	α	β	γ	δ	γ_0	Case	Class
Apatite	2.11	2.52	2.34	1.16		4.9	IV
Beryllium	2.07	1.80	3.56	0.82	4.48	4.5.1	I
Beryl	3.62	4.11	11.81	1.33	10.65	4.13.3	V
Cadmium	2.62	5.95	6.80	1.89		4.11	II
Cobalt	4.74	4.07	14.69	0.94	12.42	4.5.3	V
Ice	4.57	4.26	13.51	1.10	12.49	4.5.3	V
Hafnium	3.54	3.25	7.72	0.94	9.13	4.5.1	I
Magnesium	3.74	3.61	9.20	1.02	10.06	4.5.1	I
Rhenium	4.22	3.78	11.77	1.06	11.10	4.5.3	V
Titanium	3.88	3.47	8.31	0.75	10.03	4.5.1	I
Thallium	7.27	5.62	16.93	0.37	19.09	4.5.1	I
Yttrium	3.16	3.20	7.81	1.00	8.45	4.13.1	I
Zinc	1.57	4.17	2.40	1.69		4.9	IV

visual plotter. Figures 16–44 give the normal curve and wave front curve for each of the 13 crystals listed in Table 7. The axes for the normal curve extend from -3 to 3 with tick marks at intervals of 1. The axes for the wave front curve extend from -3 to 3 with tick marks at intervals of 1.5. In most cases the cuspidal triangles on W_+ do not show up well due to the scale used. In order to improve this situation, 'blow ups' are shown of a cusp for Cadmium, Ice and Zinc. In fig. 24 for Cadmium $\bar{z} \times \bar{y}$ magnification is 3000 $\times$ 300, with the $\bar{z}$ axis tick mark at intervals of 0.0005. In fig. 29 for Ice the magnification is 30 $\times$ 30, with tick marks at intervals of 0.1. The origin for this figure is at (0.7, 0.7). In fig. 44 for Zinc the magnification is 600 $\times$ 300 with tick marks along the $\bar{y}$ axis at intervals of 0.001.

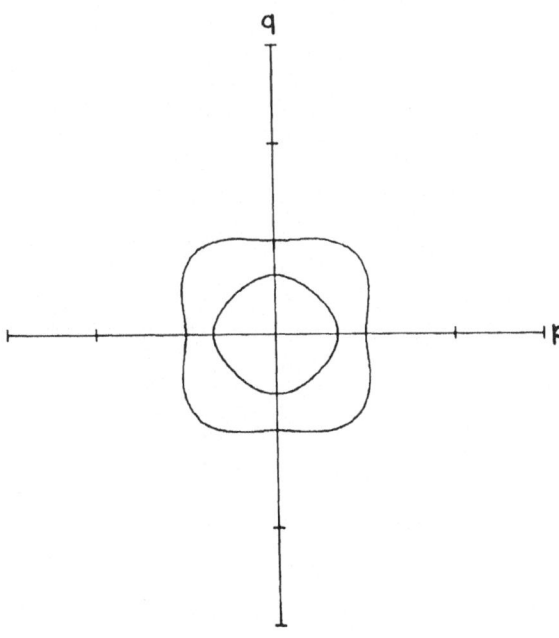

Figure 16.   Apatite normal curve.

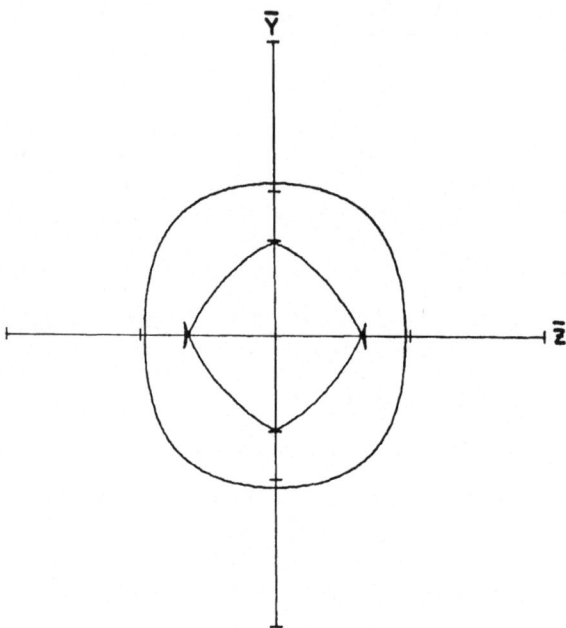

Figure 17.   Apatite wave front curve.

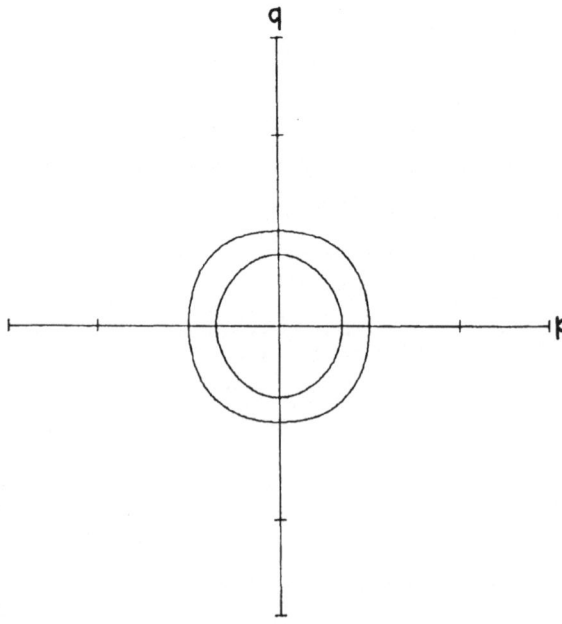

Figure 18.   Beryllium normal curve.

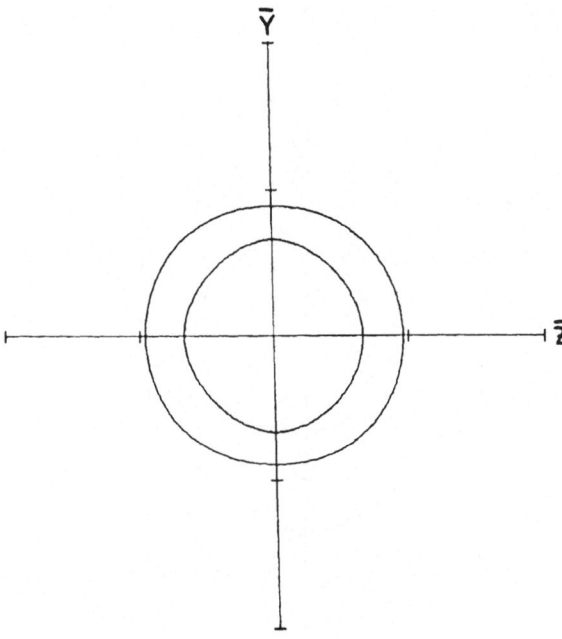

Figure 19. Beryllium wave front curve.

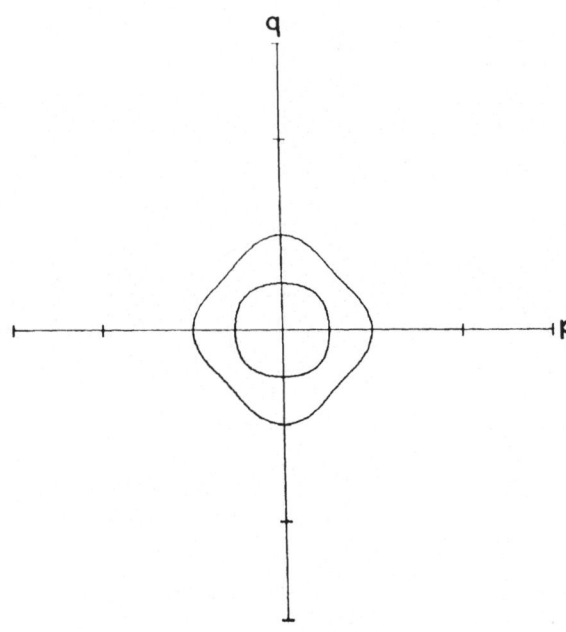

Figure 20. Beryl normal curve.

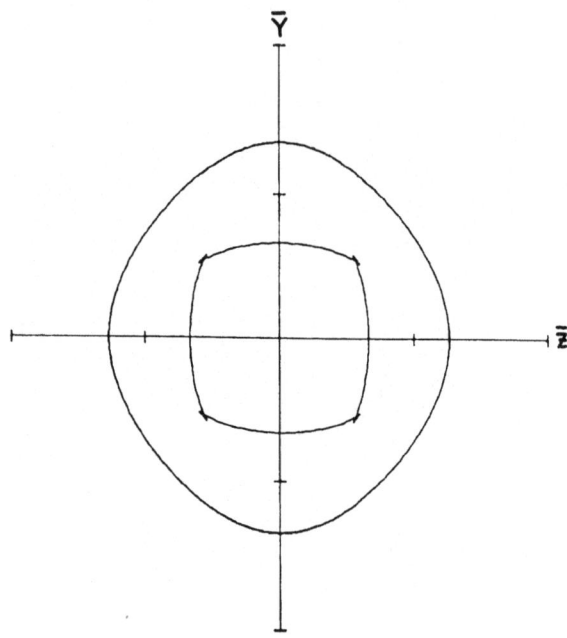

Figure 21.   Beryl wave front curve.

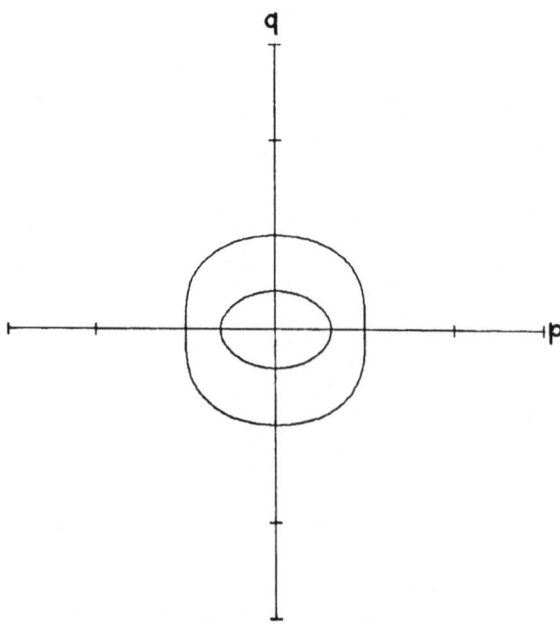

Figure 22.   Cadmium normal curve.

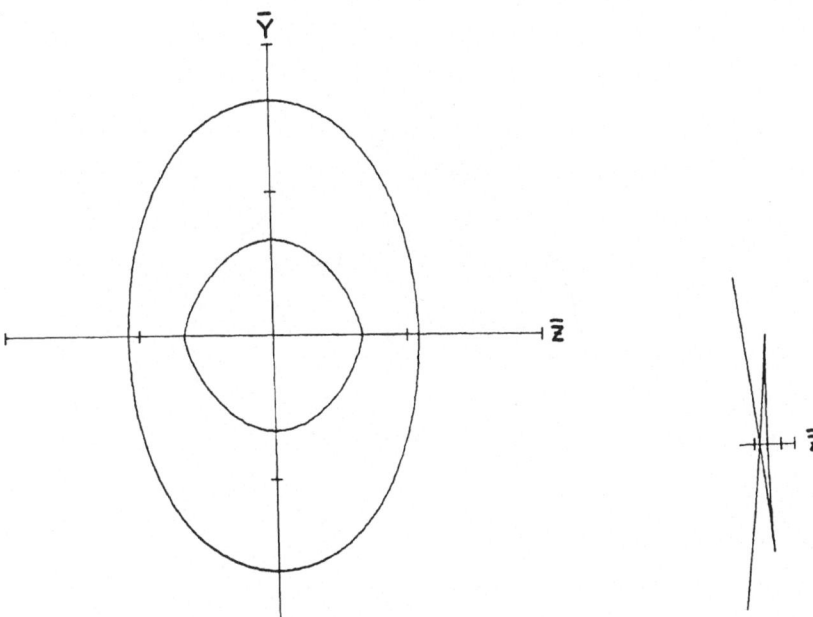

Figure 23. Cadmium wave front curve.

Figure 24. Cadmium wave front curve blow up.

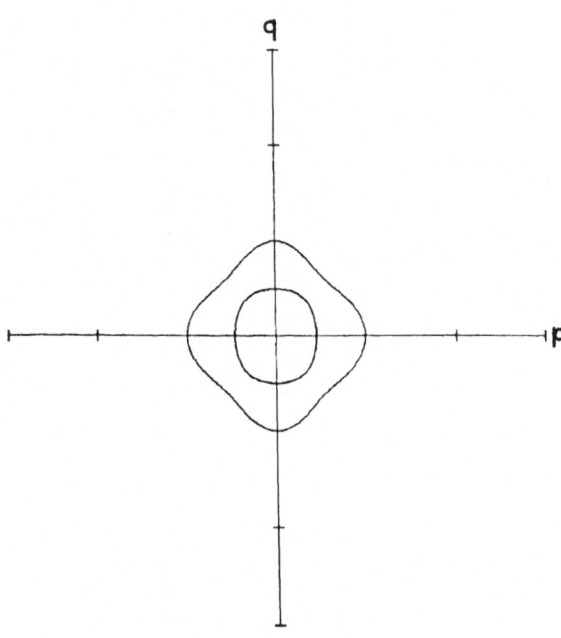

Figure 25. Cobalt normal curve.

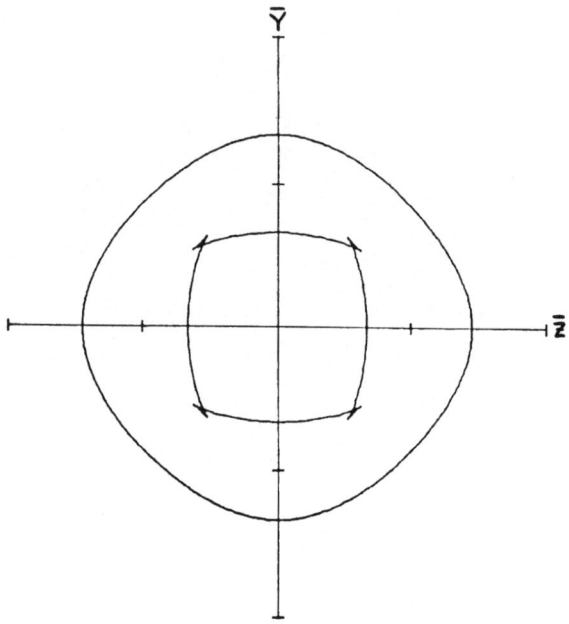

Figure 26.   Cobalt wave front curve.

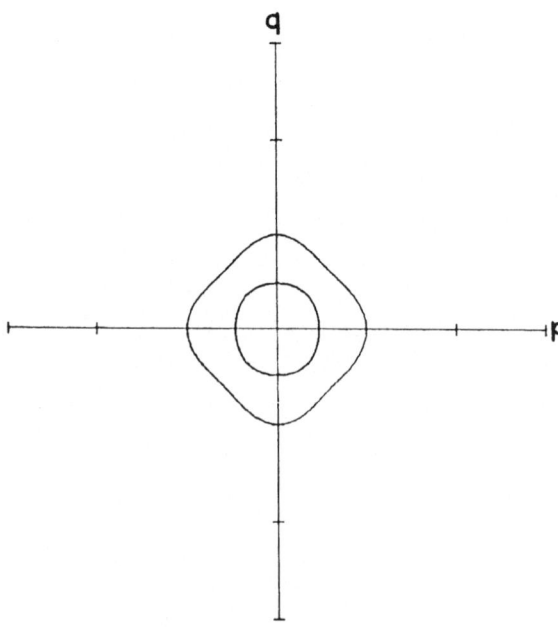

Figure 27.   Ice normal curve.

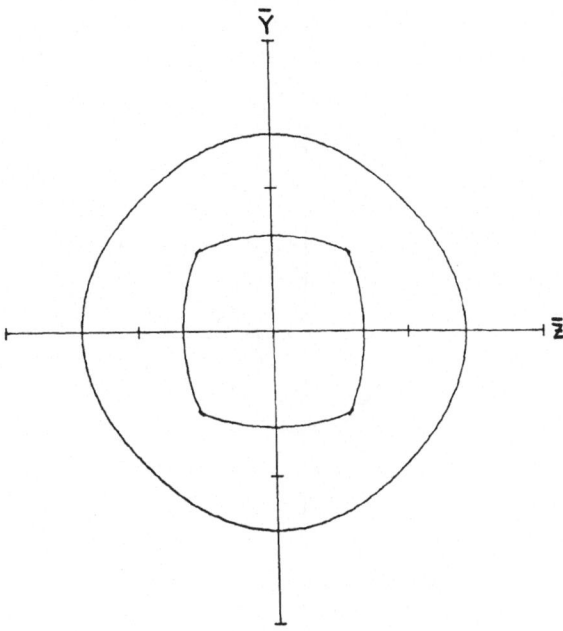

Figure 28.   Ice wave front curve.

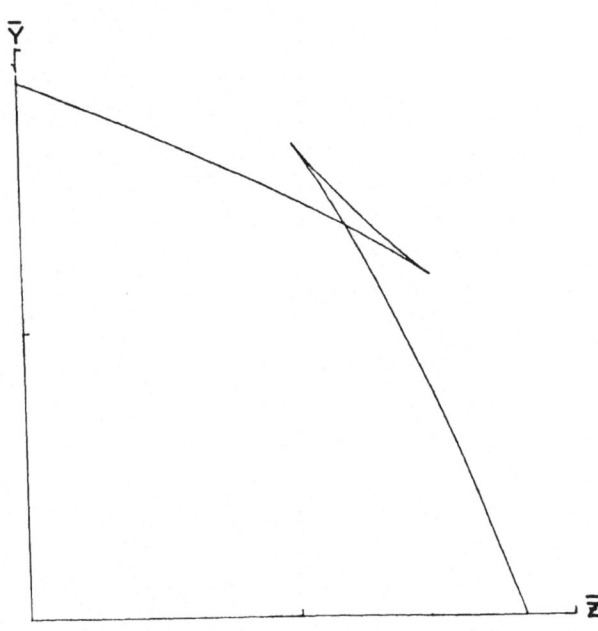

Figure 29.   Ice wave front curve blow up.

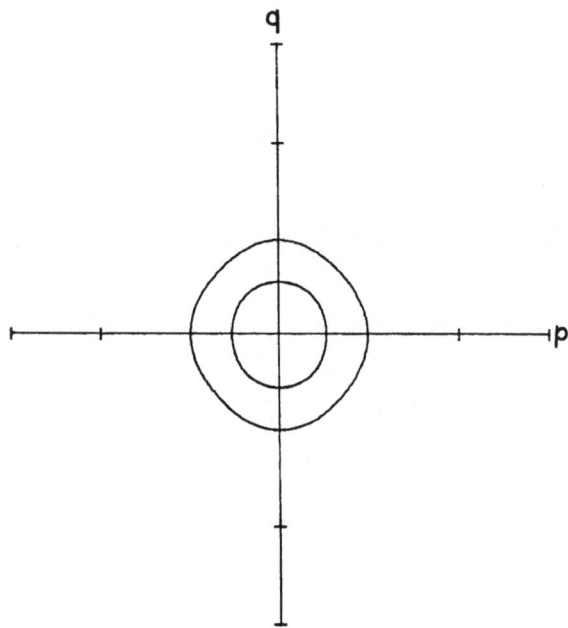

Figure 30.   Hafnium normal curve.

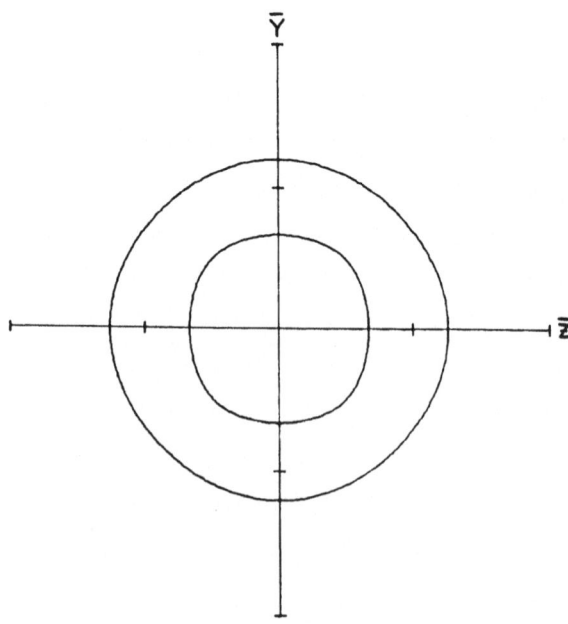

Figure 31.   Hafnium wave front curve.

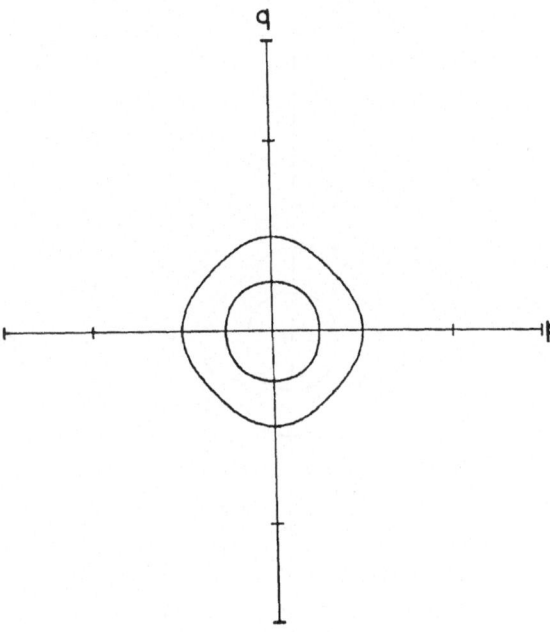

Figure 32. Magnesium normal curve.

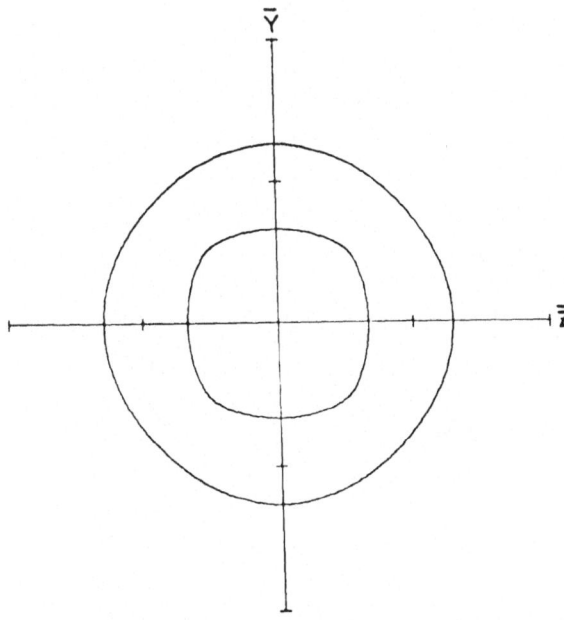

Figure 33. Magnesium wave front curve.

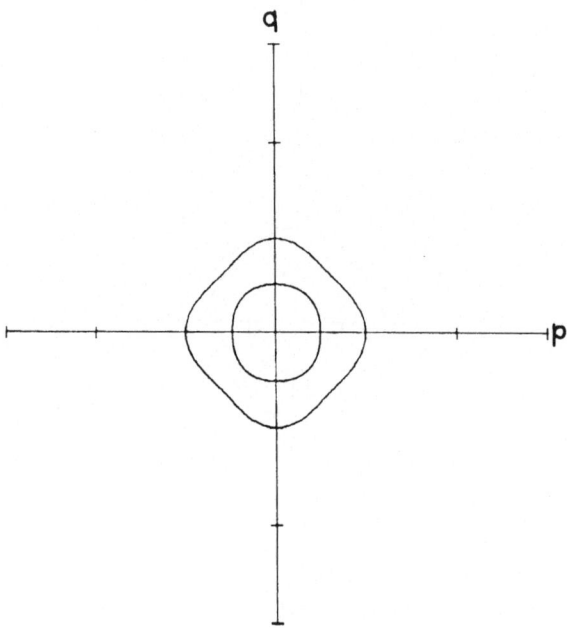

Figure 34.   Rhenium normal curve.

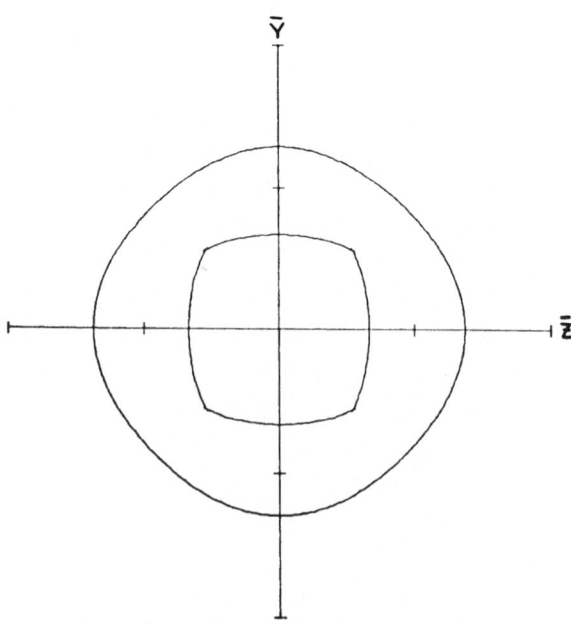

Figure 35.   Rhenium wave front curve.

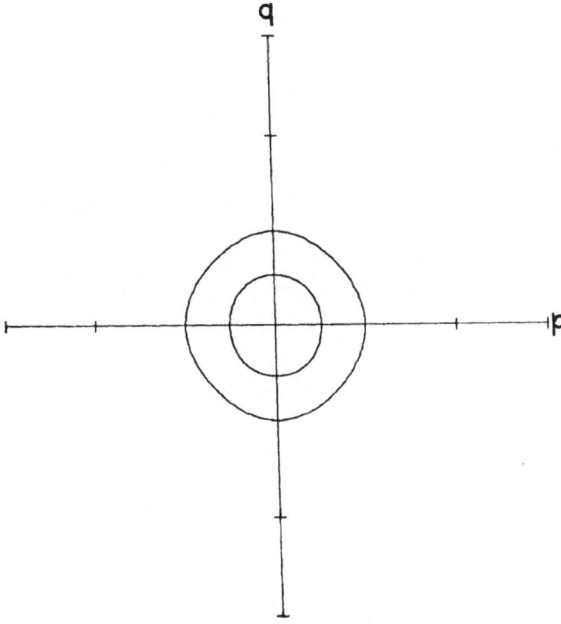

Figure 36. Titanium normal curve.

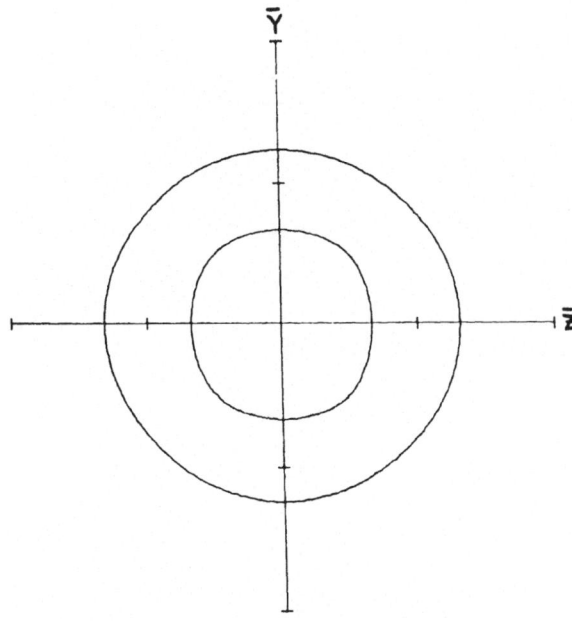

Figure 37. Titanium wave front curve.

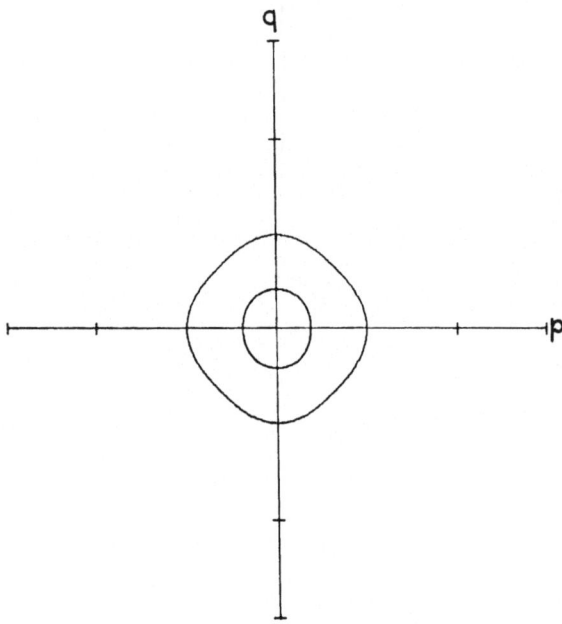

Figure 38.   Thallium normal curve.

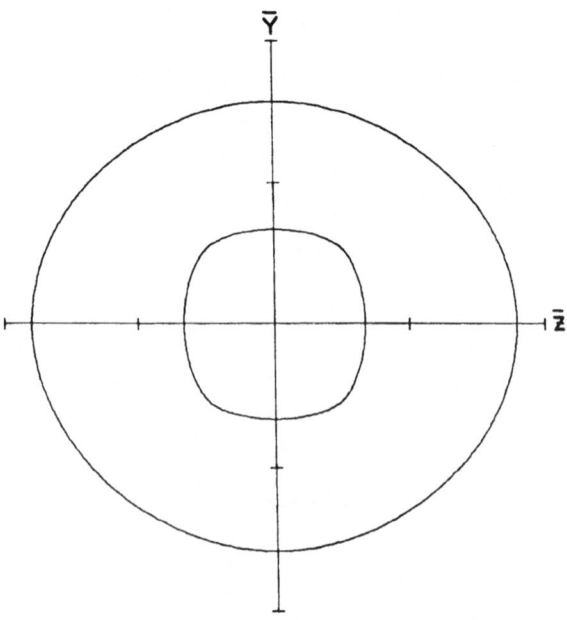

Figure 39.   Thallium wave front curve.

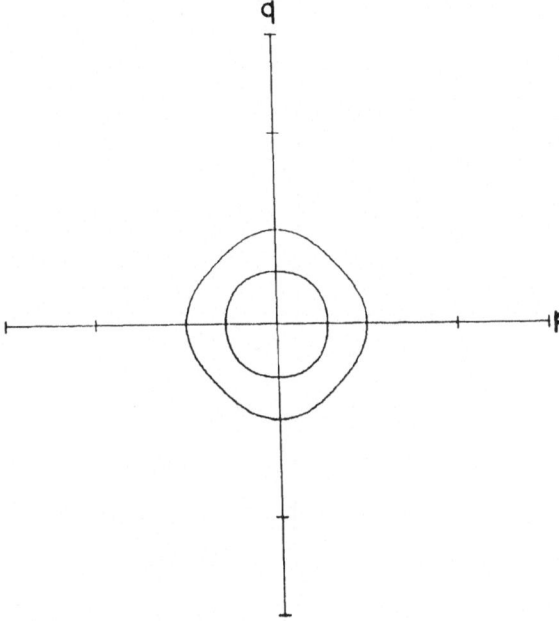

Figure 40.   Yttrium normal curve.

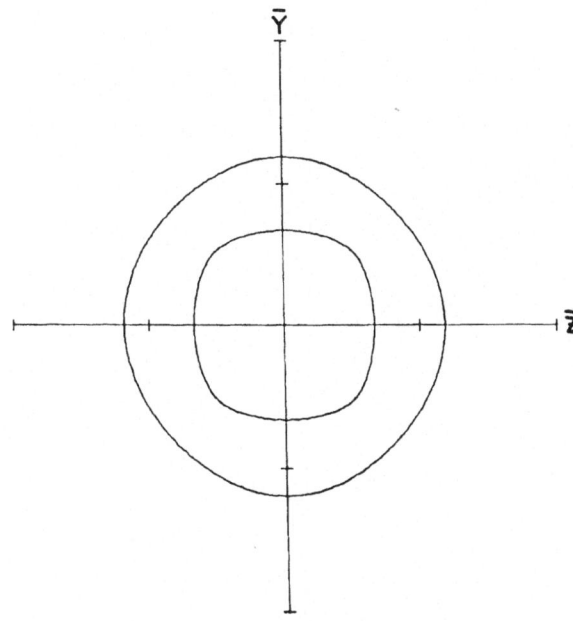

Figure 41.   Yttrium wave front curve.

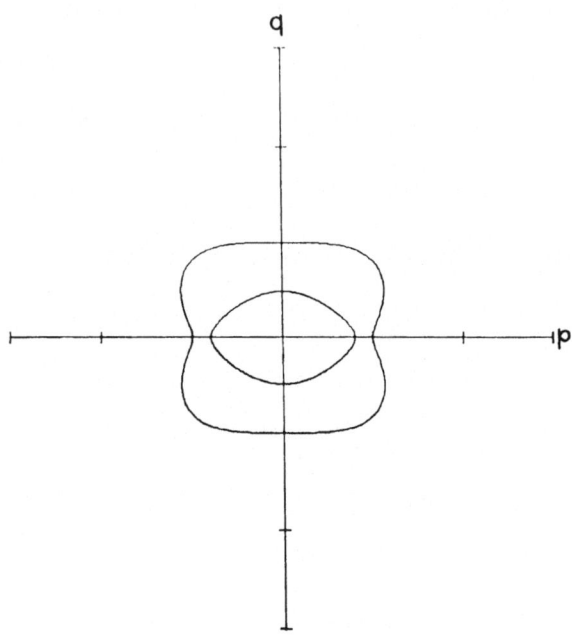

Figure 42.   Zinc normal curve.

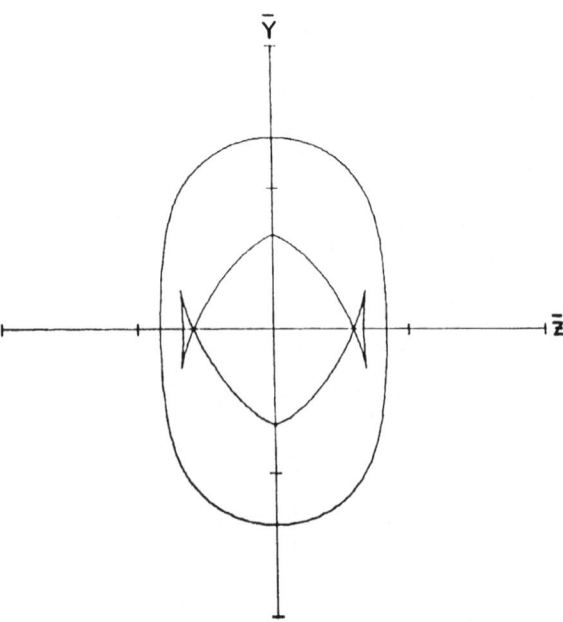

Figure 43.   Zinc wave front curve.

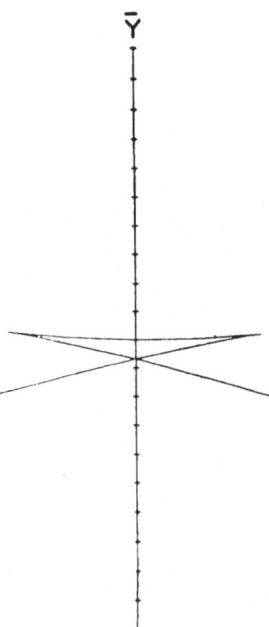

Figure 44.   Zinc wave front curve blow up.

11. Wave Front Construction When the Normal Curve has Double Points – Convex Hull

When the Q operator is strictly hyperbolic, the wave front construction given in section 2.8 is adequate. However when Q ceases to be strictly hyperbolic, so that N contains double points, this construction breaks down. As was noted in (2.5.1), a double point on N implies that $B^2(\theta) - 4A(\theta) = 0$ for certain values of θ. Since $\sqrt{B^2 - 4A}$ appears in the denominator of eqns. (2.8.8) and (2.8.9) for the parametric representation of W, the source of the trouble is obvious.

One way out of this difficulty (following Duff [21]) is to perturb the coefficients of Q in such a way that Q is now strictly hyperbolic. Let ϵ be a measure of this perturbation. Then the wave front curve, when N has double points, will be obtained as the limit of the perturbed W as ϵ tends to zero.

Figures 45–48 attempt to illustrate this perturbation for a situation in which the double points on N are caused by $\alpha = 1$. In fig. 45, $\alpha = 0.95$ ($\beta = 2$ and $\gamma = 1.5$) and it is seen that N_+ and N_- come very near to one another where they cross the p-axis. In figure 46 the (almost) vertical side of the right cuspidal triangle is very near to W_- and this is true over an extended length of W_-. When α approaches closer to 1, as in fig. 47 where $\alpha = 0.999$ ($\beta = 2$ and $\gamma = 1.6$), N_+ and N_- are so close that the figure is unable to show their separation. Figure 48 shows that the side of the cuspidal triangle on W_+ and the corresponding length of W are (almost) joined to form a vertical edge. This linear portion of W (obtained in the limit as $\alpha \to 1$) is called a *lid*

55

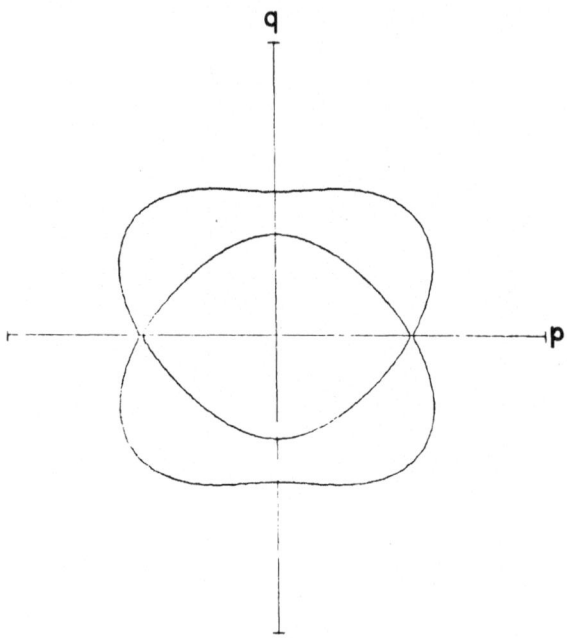

Figure 45.   Normal curve for $\alpha = 0.95, \beta = 2, \gamma = 1.5$.

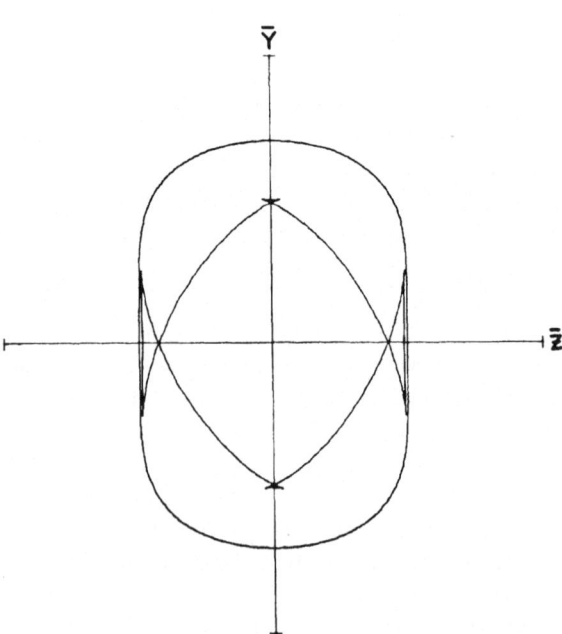

Figure 46.   Wave front curve for $\alpha = 0.95, \beta = 2, \gamma = 1.5$.

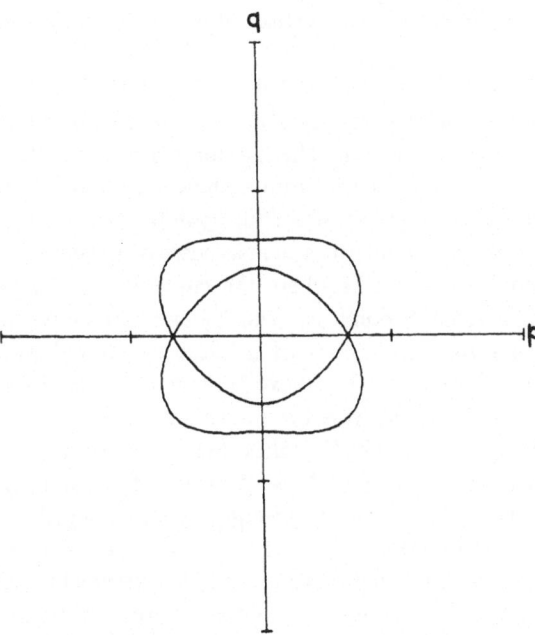

Figure 47. Normal curve for $\alpha = 0.999, \beta = 2, \gamma = 1.6$.

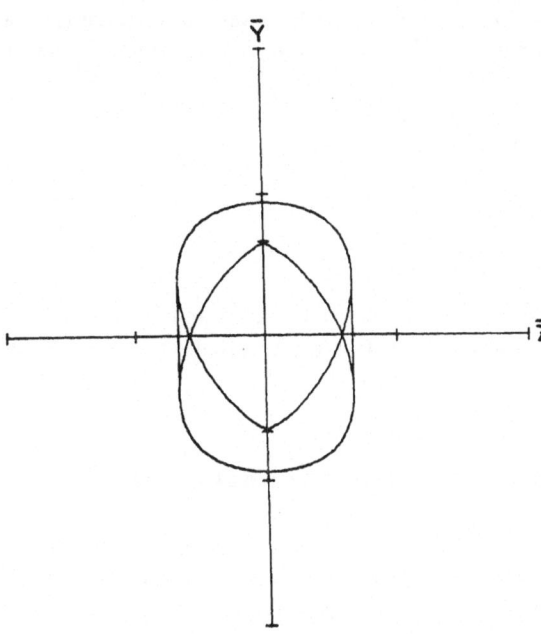

Figure 48. Wave front curve for $\alpha = 0.999, \beta = 2, \gamma = 1.6$.

and it must be adjoined to the curvilinear portion of W_- to form the *convex hull* (Courant and Hilbert [18]).

In the terminology of Integral Geometry (see Santalo [26]) the lines described by (2.8.2), whose envelope gives the wave front curve, are known as support lines and the function $V(\theta)$ is called the support function. Then, according to Courant and Hilbert [18], the general method for constructing W from a knowledge N is, 'Portions of N which are smoothly curved, of course, are mapped into points of the regular envelope of the planes polar to these points. Singularities such as cusps of the envelope must be expected at points corresponding to points where the curvature of N vanishes. If a portion of a surface N has a singular point, carrying a bundle of supporting planes, then the corresponding poles form a portion of a linear x-manifold which at the boundary is tangent to this envelope.' The above statement applies to a space of n-dimensions, so that when $n = 2$, a plane wave becomes a line wave.

Another argument (Payton [12]) which makes the need for wave front lids plausible, proceeds as follows. At a corner of N, $V'(\theta)$ is undefined and eqns. (2.8.8) and (2.8.9) (which are simply explicit solutions of eqns. (2.8.2) and (2.8.3) for $\bar{z}$ and $\bar{y}$) are indeterminate. This is easily seen to be so since, for $\alpha = 1$, $\sin\theta$ and $\sqrt{B^2 - 4A}$ simultaneously vanish. Huyghens' construction can readily be extended to handle such a situation. First note that eqn. (2.8.2) remains valid at a corner on N since $V(\theta)$ is continuous there. Next, from (2.7.16), (2.7.20) and (2.7.21), the normal curve slope can be expressed as

$$\frac{dq}{dp} = -\frac{\bar{z}}{\bar{y}} . \tag{2.11.1}$$

This equation relates the $(\bar{z}, \bar{y})$ coordinates on W to the slope (at the corresponding θ-value) on N at any smooth point on N and remain valid as a corner (say) at $\theta = \theta_0$ is approached from above or below. Now if $a \leqslant (dq/dp) \leqslant b$ at the corner on N, then from equation (2.11.1)

$$a \leqslant -\frac{\bar{y}}{\bar{z}} \leqslant b \quad \text{at} \quad \theta = \theta_0, \tag{2.11.2}$$

and from (2.8.2)

$$\bar{z}\cos\theta_0 + \bar{y}\sin\theta_0 = V(\theta_0). \tag{2.11.3}$$

Equation (2.11.3) gives a linear relation between $\bar{z}$ and $\bar{y}$ and (2.11.2) shows that this line in the $(\bar{z}, \bar{y})$ plane is of finite length. This suggests that a corner on N maps into a lid on W.

12. Necessary Condition for the Existence of Wave Front Lids

The analysis of the preceding section, based solely on properties of the differential operator Q, imply the existence of wave front lids. This general theory notwithstanding, it was noticed by Weitzner [27] when constructing the Green's function matrix for the linearized Alfvén equations of magneto-hydrodynamics, that such lids were in fact not required. The existence or nonexistence of wave front lids, in a situation where Huyghens' construction calls for their existence, is intimately connected with the form of the differential operator M appearing on the right side of the equation

$$Q\left(\frac{\partial}{\partial y}, \frac{\partial}{\partial z}, \frac{\partial}{\partial \tau}\right)[u] = M[\delta(y)\delta(z)\delta(\tau)], \tag{2.12.1}$$

where M may take either of two forms

$$M_1 = k_1 \frac{\partial^2}{\partial z^2} + k_2 \frac{\partial^2}{\partial y^2} - k_3 \frac{\partial^2}{\partial \tau^2}, \tag{2.12.2}$$

or

$$M_2 = k_4 \frac{\partial^2}{\partial z \partial y}, \tag{2.12.3}$$

and the k's are constants. (This k-notation is used in this section only, it should not be confused with k_1 and k_2 as defined in section 2.8.). With the M operator so defined, it is seen that eqn. (2.12.1) includes the four equations in (1.6.3) which describe the two-dimensional motion of a transversely isotropic elastic solid subjected to a point impulsive source.

In situations where the differential equation (2.12.1) is derivable from a (lower order) symmetric hyperbolic system of equations, the k_i coefficients of the M operators are not entirely independent. Based on the results to be stated below in Table 8, the restrictions placed on the constants k_i are such that the necessary condition for wave front lid existence is not satisfied. This is in agreement with the results found by Bazer and Yen [28].

Payton [29], by considering the integral solution to (2.12.1) obtained by Fourier-Laplace transforms, has shown that a necessary condition for the existence of lids depends upon a certain residue contribution which arises in connection with the integral evaluation. The details are here omitted, but the results are summarized in Table 8. The cases refer to those of section 2.5, for which N has multiple points.

As an example of the application of Table 8, consider the question of the existence of wave front lids for the equation (from (1.6.3))

$$Q[w_3] = -K[\delta(y)\delta(z)\delta(\tau)], \tag{2.12.4}$$

Table 8.   Necessary Conditions for the Existence of Wave Front Lids

Case	$Q[u] = M_1[\delta(y)\delta(z)\delta(\tau)]$	$Q[u] = M_2[\delta(y)\delta(z)\delta(\tau)]$
5	$k_1(1-\beta) + k_2(1-\alpha) + k_3(\alpha\beta - 1) \neq 0$	$k_4 \neq 0$
6	$k_1(1-\beta) + k_2(1-\alpha) + k_3(\alpha\beta - 1) \neq 0$	$k_4 \neq 0$
7	$k_1 \neq k_3$	no lid possible
8	$k_2 \neq k_3$	no lid possible
9	$k_1 \neq k_3$ and $k_2 \neq k_3$	no lid possible
10	no lid possible	no lid possible

where K is given by (1.6.7) as

$$K = \frac{\partial^2}{\partial z^2} + \beta \frac{\partial^2}{\partial y^2} - \frac{\partial^2}{\partial \tau^2}. \tag{2.12.5}$$

Comparing (2.12.5) and (2.12.2) gives

$$k_1 = 1, \qquad k_2 = \beta \quad \text{and} \quad k_3 = 1.$$

Then, if (say) Case 5) is treated,

$$k_1(1-\beta) + k_2(1-\alpha) + k_3(\alpha\beta - 1) = (1-\beta) + \beta(1-\alpha) + (\alpha\beta - 1) = 0,$$

so that the above necessary condition for lid existence is not satisfied. A similar calculation shows that *wave front lids do not exist for any of the equations (1.6.3).*

The reader may jump to the conclusion that lids do not exist for any anisotropic elastic wave problems. Such is not the case, as has been demonstrated by Burridge [30] for cubic crystals in 3-space dimensions.

13. The Wave Front Shape when the Normal Curve has Double Points – Conclusion

Based on the discussion of sections 2.11 and 2.12, it is now clear that the wave front curve associated with the free space solutions to (1.6.3) can be obtained from the line wave envelope construction. When the normal curve contains double points, the wave front lids are simply deleted. Figures 49–51 show three examples of such wave fronts. The normal curves for these figures were given previously in figs. 6–8.

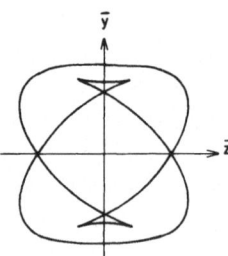

Figure 49.   Wave front curve for case 7.1. ($\alpha = 1, \beta = 0.64, \gamma = 0.42$)

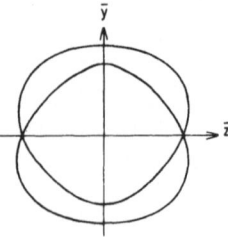

Figure 50.   Wave front curve for case 7.2. ($\alpha = 1, \beta = 0.64, \gamma = 1.28$)

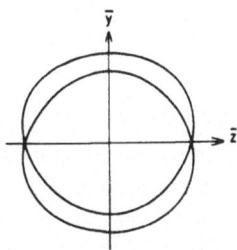

Figure 51. Wave front curve for case 7.3. ($\alpha = 1, \beta = 0.64, \gamma = 1.50$)

14. Remarks on the Wave Front Shape when the Line Load is Not Perpendicular to the Material Symmetry Axis

The 3-dimensional version of the plane strain problem governed by eqns. (1.6.3) has a uniform impulsive line load along the x- (or x_1) axis. Suppose now that a new problem is considered in which the loading axis is not normal to the material symmetry axis. Let the coordinate frame $Oxyz$ or $Ox_1x_2x_3$ be a coordinate frame with the x_3-axis taken as the symmetry axis. Let $Ox_1'x_2'x_3'$ be a second coordinate frame such that the x_1'-axis coincides with the line load. With respect to the primed coordinate system the applied load (per unit mass) is

$$f_i' = (0, a_2', a_3') \, \delta(x_2') \delta(x_3') \, \delta(t). \tag{2.14.1}$$

The equations of motion, in the unprimed frame are, by (1.1.6)

$$c_{ijkl} \frac{\partial^2 u_k}{\partial x_l \partial x_j} + \rho f_i = \rho \frac{\partial^2 u_i}{\partial t^2}, \tag{2.14.2}$$

where, using (1.2.1),

$$f_i = (\alpha_{21}a_2' + \alpha_{31}a_3', \, \alpha_{22}a_2' + \alpha_{32}a_3', \, \alpha_{23}a_2' + \alpha_{33}a_3') \delta(x_2') \delta(x_3') \delta(t). \tag{2.14.3}$$

The loading for the equations of motion (2.14.2), in an obvious notation, thus has the form

$$f_i = (a_1, a_2, a_3) f(x_1, x_2, x_3, t), \tag{2.14.4}$$

where

$$f(x_1, x_2, x_3, t) = \delta(x_2') \delta(x_3') \delta(t), \tag{2.14.5}$$

$$x' = \alpha_{21}x_1 + \alpha_{22}x_2 + \alpha_{23}x_3, \tag{2.14.6}$$

and

$$x_3' = \alpha_{31}x_1 + \alpha_{32}x_2 + \alpha_{33}x_3. \tag{2.14.7}$$

The above problem now has precisely the form of the 3-dimensional problem whose equations were uncoupled in (1.7.3). However now the left side differential operator will (in general) be a sixth order operator, the product of Q and a second order operator L. Suppose a solution (e.g., for u_1) of the form

$$u_1(x_1, x_2, x_3, t) = \bar{u}_1(x'_2, x'_3, t), \tag{2.14.8}$$

is sought. Then using $\partial u_1/\partial x_i = (\partial \bar{u}_1/\partial x'_k)(\partial x'_k/\partial x_i) = \alpha_{ki}\partial \bar{u}_1/\partial x'_k$ etc., the problem again reduces to one of plane strain in the variables x'_2, x'_3 and t. Using the primed variables causes the L and Q operators to become more extended in that they will now contain a richer variety of mixed derivatives, but Q will remain a fourth order, linear, homogeneous, partial differential operator with constant coefficients.

Thus the line load problem of this section is again amenable to the wave front construction technique of this chapter. There is a major difference, however, between the normal and wave front curves previously considered and those associated with a line load not perpendicular to the symmetry axis. In the present case N and W will each contain three branches.

PART II.   THREE SPACE DIMENSIONS

In the remainder of this chapter, the normal and wave front surfaces associated with a three dimensional point source will be treated. Since much of the analysis closely parallels that given in the first part of this chapter for the two dimensional problem, it will be possible to state the results more briefly. The main results are stated in sections 16 and 18, wherein it is shown that the normal and wave surfaces are obtained as surfaces of revolution by revolving the corresponding two dimensional normal and wave front curves about the symmetry axis. One additional sheet must be adjoined, due to the operator L.

15. The Characteristic Partial Differential Equation for the Wave Front

For an initial value problem governed by eqns. (1.7.3) in which the disturbance is concentrated at the origin in 3-space, a wave front will form at $\tau = 0$ and propagate away from the origin as time increases. Let the equation of the wave front be

$$W: \chi(x, y, z, \tau) = 0. \tag{2.15.1}$$

Then, since χ is a characteristic surface, it must satisfy

$$L(\chi_x, \chi_y, \chi_z, \chi_\tau)Q(\chi_x, \chi_y, \chi_z, \chi_\tau) = 0 \quad \text{on} \quad \chi(x, y, z, \tau) = 0. \tag{2.15.2}$$

Equation (2.15.2) is the characteristic partial differential equation for the wave front W associated with the three dimensional Green's function problem formulated in section (1.7). For certain of the displacement components (u_3, v_3, w_1, w_2 and w_3) the L operator on the left side of (2.15.2) will not be present.

Assuming $\chi_\tau \neq 0$, a solution to (2.15.2) may be sought in the form

$$\chi(x, y, z, \tau) = \psi(x, y, z) - \tau = 0. \tag{2.15.3}$$

Substituting (2.15.3) into (2.15.2) gives the first order nonlinear partial differential equation for ψ

$$L(\psi_x, \psi_y, \psi_z, -1)Q(\psi_x, \psi_y, \psi_z, -1) = 0. \tag{2.15.4}$$

16. The Normal Surface – Rotational Symmetry

At any instant of time $\tau > 0$, the wave front surface

$$W: \psi(x, y, z) - \tau = 0, \tag{2.16.1}$$

has a normal vector with components $\psi_x \equiv q_1$, $\psi_y \equiv q_2$ and $\psi_z \equiv p$ which must satisfy the sixth degree algebraic equation

$$N: L(q_1, q_2, p, -1)Q(q_1, q_2, p, -1) = 0, \tag{2.16.2}$$

where the 3-space operators L and Q have been previously defined in (1.5.20) and (1.5.21). In (q_1, q_2, p) space the surface $LQ = 0$ gives the surface of wave normals and will be referred to as the *normal surface* or slowness surface.

Introduce spherical polar coordinates (R, θ, ϕ) with the p-axis (symmetry axis) as the polar axis, by

$$q_1 = R \sin \theta \cos \phi, \qquad q_2 = R \sin \theta \sin \phi \quad \text{and} \quad p = R \cos \theta, \tag{2.16.3}$$

where $R \geqslant 0, 0 \leqslant \theta \leqslant \pi$ and $0 \leqslant \phi < 2\pi$. Then eqn. (2.16.2) becomes

$$[R^2(\cos^2 \theta + \delta \sin^2 \theta) - 1] \, [AR^4 - BR^2 + 1] = 0, \tag{2.16.4}$$

where

$$A(\theta) = \alpha \cos^4 \theta + \gamma \cos^2 \theta \sin^2 \theta + \beta \sin^4 \theta, \tag{2.16.5}$$

and

$$B(\theta) = (\alpha + 1)\cos^2 \theta + (\beta + 1) \sin^2 \theta. \tag{2.16.6}$$

(Note that, although eqns. (2.16.5) and (2.16.6) formally agree with (1.5.26) and (1.5.27), the interpretation of θ is different.)

It is apparent from eqn. (2.16.4), since ϕ does not appear there, that the normal surface is a figure of revolution about the p-axis. The normal surface is composed of three sheets, with parametric (θ, ϕ) representation.

$$N_0: q_1 = R_0(\theta) \sin \theta \cos \phi, \qquad q_2 = R_0(\theta) \sin \theta \sin \phi,$$

$$p = R_0(\theta) \cos \theta, \tag{2.16.7}$$

and

$$N_\pm: q_1 = R_\pm(\theta) \sin \theta \cos \phi, \qquad q_2 = R_\pm(\theta) \sin \theta \sin \phi,$$

$$p = R_\pm(\theta) \cos \theta. \tag{2.16.8}$$

In the above

$$R_0(\theta) = [\cos^2 \theta + \delta \sin^2 \theta]^{-1/2}, \tag{2.16.9}$$

and

$$R_\pm(\theta) = \left[\frac{B(\theta) \pm \sqrt{B^2(\theta) - 4A(\theta)}}{2A(\theta)}\right]^{1/2} \tag{2.16.10}$$

By (2.16.10) $R_+ \geqslant R_-$ so that the sheet N_- is always contained within the sheet N_+. Of course N_+ and N_- are simply figures of revolution obtained by rotating the plane curves (2.2.8) about the p-axis. For this reason it is sufficient to discuss the plane curve (normal curve) in the $p, q \ (= \sqrt{q_1^2 + q_2^2})$ plane in order to classify the normal surface shape. But since the geometry of this curve has been discussed earlier at some length, the classification as to shape of the normal surface sheets $N_\pm$ follows from section 2.6 and the tables therein.

The sheet N_0 is simply an ellipsoid obtained by rotating the plane ellipse

$$p^2 + \delta q^2 = 1, \tag{2.16.11}$$

about the p-axis.

17. Multiple Points of the Normal Surface

In three dimensions there can exist a variety of multiple points on the normal surface. For example if $\alpha = 1$ then the point $(0, 0, 1)$ in (q_1, q_2, p) space is a triple point, while for $\beta = 1$, there will be a circle of double points on $q_1^2 + q_2^2 = 1, p = 0$ (provided $\delta \neq 1$, if $\delta = 1$ it is a circle of triple points). The location of multiple points connected solely with sheets $N_\pm$ follows easily from Cases 5)–10) of section 2.5 and the fact that N is a surface of revolution.

It remains now to explore the possible intersections of the normal curve ellipse

$$N_0 : p^2 + \delta q^2 = 1, \tag{2.17.1}$$

with the other two branches of the normal curve N_- and N_+. Now on the p-axis $(q_1 = q_2 = 0)$, eqn. (2.16.2) gives

$$Q(0, 0, p, -1) = 0, \tag{2.17.2}$$

or

$$\alpha p^4 - (\alpha + 1)p^2 + 1 = (\alpha p^2 - 1)(p^2 - 1) = 0, \tag{2.17.3}$$

and

$$L(0, 0, p, -1) = 0 \quad \text{or} \quad p^2 - 1 = 0. \tag{2.17.4}$$

Clearly the normal surface has double points on the p-axis (axis of symmetry) at $p = \pm 1$ where the sheet N_0 touches one (except for Cases 7) and 9) where it intersects both) of the sheets N_- or N_+. This statement remains true regardless of the relative values of the elasticity parameters α, β, γ and δ. Thus the normal surface for a transversely isotropic solid always has (at least) two multiple points. However, except when $\alpha = 1$, the point $(0, 0, 1)$ of N is not a corner point, since the sheets of N are smooth there and do not cross.

In order to ascertain whether or not the sheet N_0 makes contact with $N_\pm$ off the

p-axis (following Payton [14]) it is helpful to examine the local behavior of N_0 and $N_\pm$ in the neighborhood of the point $(0, 0, 1)$. Since most hexagonal crystals fall under Case 4), further examination of possible self intersection points for N will be restricted to this case. The extension to the remaining nine cases is straightforward.

Near the point $(1, 0)$ in the p, q plane, the θ parametric behaviour of N_0 and N_+ may be deduced from

$$p_0(\theta) = 1 - \frac{\delta}{2}\theta^2 + O(\theta^4) \qquad (2.17.5)$$

and

$$p_+(\theta) = 1 - \frac{\gamma - \beta - 1}{2(\alpha - 1)}\theta^2 + O(\theta^4). \qquad (2.17.6)$$

Thus if $-2\sqrt{\alpha\beta} < \gamma \leqslant (\beta + 1)$ then locally (near the p-axis) the sheet N_0 lies inside the sheet N_+. While if $(\beta + 1) < \gamma < (1 + \alpha\beta)$, N_0 may or may not lie inside N_+, depending on the relation of δ to $(\gamma - \beta - 1)/(\alpha - 1)$. The information contained in eqns. (2.17.5) and (2.17.6), together with the intersection points of the normal curve branches $N_\pm$ and N_0 with the q-axis, is sufficient to determine the existence of contact points between N_0 and $N_\pm$ off the p-axis.

Table 9 summarizes the conditions of the elasticity parameters α, β, γ and δ for the existence of off p-axis contact points between N_0 and $N_\pm$. Since the normal curve possesses symmetry about the p and q-axes, only information pertaining to the first quadrant of the p, q plane will be given. Thus for every contact point between N_0 and $N_\pm$ in the first quadrant of the normal plane curve, there will exist three additional (symmetrically located) contact points. When the normal curve is rotated to form the normal surface, these four contact points give rise to two circles of contact between one of the sheets N_- or N_+ and N_0.

Table 9. Contact Points Between Branches of the Normal Curve N_0 and $N_\pm$ Which Lie in the First Quadrant of the p, q Plane, but off the p-axis, $(\alpha > 1$ and $\beta > 1)$

γ-range	δ-range	Contact between N_0 and $N_\pm$
$-2\sqrt{\alpha\beta} < \gamma \leqslant (\beta + 1)$	$0 < \delta \leqslant 1$	N_0 contacts N_+
	$1 < \delta < \beta$	N_0 does not contact $N_\pm$
	$\beta \leqslant \delta$	N_0 contacts N_-
$(\beta + 1) < \gamma < (\alpha + \beta)$	$0 < \delta \leqslant (\gamma - \beta - 1)/(\alpha - 1)$	N_0 does not contact $N_\pm$
	$(\gamma - \beta - 1)/(\alpha - 1) < \delta \leqslant 1$	N_0 contacts N_+
	$1 < \delta < \beta$	N_0 does not contact $N_\pm$
	$\beta \leqslant \delta$	N_0 contacts N_-
$\gamma = (\alpha + \beta)$	$0 < \delta < 1$	N_0 does not contact $N_\pm$
	$\delta = 1$	N_0 coincides with N_+
	$1 < \delta < \beta$	N_0 does not contact $N_\pm$
	$\beta \leqslant \delta$	N_0 contacts N_-
$(\alpha + \beta) < \gamma < (1 + \alpha\beta)$	$0 < \delta < 1$	N_0 does not contact $N_\pm$
	$1 \leqslant \delta < (\gamma - \beta - 1)/(\alpha - 1)$	N_0 contacts N_+
	$(\gamma - \beta - 1)/(\gamma - 1) \leqslant \delta < \beta$	N_0 does not contact $N_\pm$
	$\beta \leqslant \delta$	N_0 contacts N_-

18. Wave Front Construction as Envelope of Plane Waves

According to the wave front construction method of Huyghens, the general form of the wave front emitted by a concentrated source at the origin in (x, y, z) space at time $\tau = 0$, is given by the envelope of the two parameters (θ, ϕ) family of plane waves

$$x \sin \theta \cos \phi + y \sin \theta \sin \phi + z \cos \theta = V(\theta)\tau, \tag{2.18.1}$$

where $1/V(\theta)$ is equal to $R_0(\theta)$, $R_-(\theta)$ or $R_+(\theta)$ depending on the wave front sheet in question. The general parametric representation of any sheet of the wave front surface is given by

$$x/\tau = \cos \phi [V(\theta) \sin \theta + V'(\theta) \cos \theta], \tag{2.18.2}$$

$$x/\tau = \sin \phi [V(\theta) \sin \theta + V'(\theta) \cos \theta], \tag{2.18.3}$$

and

$$z/\tau = [V(\theta) \cos \theta - V'(\theta) \sin \theta]. \tag{2.18.4}$$

In particular, associating the wave front sheets W_0 and $W_\pm$ with the normal surface sheets N_0 and $N_\pm$ gives

$$W_0 : x/\tau = \delta R_0 \sin \theta \cos \phi, \qquad y/\tau = \delta R_0 \sin \theta \sin \phi, \qquad z/\tau = R_0 \cos \theta, \tag{2.18.5}$$

and

$$W_\pm : \frac{x}{\tau \cos \phi} = \frac{y}{\tau \sin \phi} =$$

$$\frac{\sin \theta}{2AR_\pm} \left[2\beta \sin^2 \theta + \gamma \cos^2 \theta \mp \frac{\cos^2 \theta (k_1 \cos^2 \theta - k_2 \sin^2 \theta)}{\sqrt{B^2 - 4A}} \right], \tag{2.18.6}$$

$$z/\tau = \frac{\cos \theta}{2AR_\pm} \left[2\alpha \cos^2 \theta + \gamma \sin^2 \theta \pm \frac{\sin^2 \theta (k_1 \cos^2 \theta - k_2 \sin^2 \theta)}{\sqrt{B^2 - 4A}} \right] \tag{2.18.7}$$

where $k_1 = 2\alpha(\beta + 1) - \gamma(\alpha + 1)$ and $k_2 = 2\beta(\alpha + 1) - \gamma(\beta + 1)$.

From eqns. (2.18.5), (2.18.6) and (2.18.7) it is evident that the wave front is composed of three sheets which are figures of revolution about the z-axis. W_0 is an ellipsoid, W_- has a convex shape, while W_+ can be rather complicated in shape. Depending on the relative values of the parameters α, β and γ, W_+ may contain cuspidal edges. Cross sections of typical wave front shapes for the sheets $W_\pm$ which fall within classes I–V are shown in figs. 11–15.

19. The Existence of Wave Front Lids

As discussed earlier in section 2.11, the wave front construction method of Huyghens breaks down when the normal surface has multiple points due to self intersections (e.g., when $\alpha = 1$). In such a case the theory of first order partial differential equations calls for a flat disc or conical frustum (rotational lid) to be adjoined to the wave front sheets so as to render the resulting hull convex. However, if attention is confirmed to Case 4) which is of most physical interest, it is seen from eqns. (2.18.2)–(2.18.4) that the W construction is well defined at all points since $V'(\theta)$ is continuous. Stated another way, a corner in N caused by (say) the intersection of N_0 and N_+ does not cause a jump in $V(\theta)$ at that point. For this reason it is conjectured that no wave front lids are necessary.

20. Normal and Wave Surfaces for some Hexagonal Materials

If the plane curve

$$p^2 + \delta q^2 = 1, \qquad\qquad (2.20.1)$$

is adjoined to the plane normal curves for the hexagonal materials presented in section 2.10 and the resulting three branches are rotated about the p-axis, then the normal surface is obtained. Similarly if the plane curve

$$\bar{y}^2/\delta + \bar{z}^2 = 1, \qquad\qquad (2.20.2)$$

is adjoined to the wave front curves depicted in section 2.10 and the three branches are then rotated about the $\bar{z}$-axis, then the wave front surface is obtained. Since both of the curves (2.20.1) and (2.20.2) are ellipses, the reader's imagination can easily supply the composite figures. The shape classification of N_+ and W_+ remains the same as that given in Table 7. Musgrave [20] gives two figures which show all three branches of N and W for Beryl and Zinc. In Table 10, the results given in Table 9 are applied to some hexagonal crystals in order to establish the existence of contact curves between the sheets N_0 and N_+ or N_- at points off the p-axis. In view of the relative simplicity of the plane curves which generate N, it would not be difficult to find the coordinates of these points in the p-q plane.

Table 10. Contact Points between the sheets N_0 and N_+ or N_- off the Symmetry Axis.

Crystal	Does N_0 Contact N_-?	Does N_0 Contact N_+?
Apatite	NO	NO
Beryllium	NO	YES
Beryl	NO	YES
Cadmium	NO	NO
Cobalt	NO	NO
Ice	NO	YES
Hafnium	NO	NO
Magnesium	NO	YES
Rhenium	NO	YES
Titanium	NO	NO
Thallium	NO	NO
Yttrium	NO	YES
Zinc	NO	NO

3

Green's tensor for the displacement field in unbounded media

In this chapter a solution is constructed, by means of integral transforms to the problem of determining the displacement field behind the expanding wavefront excited by a point body force in two and three-dimensions. By applying an integration technique of Burridge [31], an explicit solution is found in the two-dimensional case. In the three-dimensional situation, with the exception of a special combination of the elastic parameters ($\gamma = \alpha + \beta$), explicit results can be obtained only on the symmetry axis. Seminal work on the subject matter of this chapter, especially on the existence of lacunas, has been done by Petrowski [32]. The survey article by Duff [24] also touches on many of the topics of this chapter and cites a number of recent references.

PART I: TWO SPACE DIMENSIONS

1. Integral Transform Representation of the Displacement Field

The central problem of Part I concerns the response of an unbounded, transversely isotropic, elastic solid to a time dependent point (i.e., line in 3-dimensions) load. The problem is one of plane strain. Major attention is directed to the situation in which the applied line load is perpendicular to the axis of material symmetry, but the solution technique can also handle cases in which the line load has an arbitrary orientation with respect to the symmetry axis (see Section 8, below).

The statement and mathematical formulation of this problem have previously been given in Section 6 of Chapter 1. Before proceeding it should be noted that if the impulsive body force has the form

$$f_i = \delta_{ip}\delta(y)\delta(z)\delta(t) \tag{3.1.1}$$

and the corresponding displacement vector u_i satisfying equations (1.1.6) is written as

$$u_i = u_{ip} \tag{3.1.2}$$

then u_{ip} is called the Green's tensor and its components satisfy

$$c_{ijkl}\frac{\partial^2 u_{kp}}{\partial x_l \partial x_j} + \rho\delta_{ip}\delta(x_2)\delta(x_3)\delta(t) = \rho\frac{\partial^2 u_{ip}}{\partial t^2}, \tag{3.1.3}$$

where $i, j, \ldots, p$ are now restricted to range over the integers 2 and 3. Thus, in the notation of Section 1.6,

$$(u_{22}, u_{23} = u_{32}, u_{33}) = \sqrt{\frac{\rho}{c_{44}}} (v_2, v_3 = w_2, w_3). \qquad (3.1.4)$$

A solution to (1.6.3) subject to zero initial conditions in free space will now be constructed with the help of integral transforms. It is appropriate here to use a double Fourier exponential transform on the space variables y and z and a Laplace transform on the time variable τ. Thus a solution to (1.6.3), e.g. for v_2 is sought in the form

$$v_2(y, z, \tau) = \frac{1}{(2\pi)^3 i} \int_C \int_{-\infty}^{+\infty} \int_{-\infty}^{+\infty} \bar{v}_2(\eta, \zeta, s) \exp\left(i(\eta y + \zeta z) + s\tau\right) d\eta d\zeta ds,$$

$$(3.1.5)$$

where

$$\bar{v}_2(\eta, \zeta, s) = \int_0^\infty \int_{-\infty}^{+\infty} \int_{-\infty}^{+\infty} v_2(y, z, \tau) \exp\left(-i(\eta y + \zeta z) - s\tau\right) dy dz d\tau. \qquad (3.1.6)$$

Applying the triple transform indicated in (3.1.6) to (1.6.3), integrating by parts, etc., gives the following algebraic expression for the transformed v_2 displacement

$$\bar{v}_2(\eta, \zeta, s) = -\frac{N(i\eta, i\zeta, s)}{Q(i\eta, i\zeta, s)}, \qquad (3.1.7)$$

with similar expressions for $\bar{w}_2$ and $\bar{w}_3$. Note that, from (1.6.5) and (1.6.6), the polynomials N and Q are homogeneous in the variables η, ζ and s of degree 2 and 4 respectively.

2. Transform Inversion and Reduction to a Residue Calculation

The triple integral (3.1.5) with $\bar{v}_2$ given by (3.1.7) will now be evaluated. It is a remarkable fact that these integrations can be performed exactly and the result displayed in closed form.

Introduce polar coorinates (r, θ) by

$$\zeta = r \cos \theta, \qquad \eta = r \sin \theta \quad \text{for} \quad 0 \leqslant r < \infty, \qquad 0 \leqslant \theta < 2\pi.$$

$$(3.2.1)$$

Equation (3.1.5), with the help of (3.1.7), may then be written as

$$v_2 = \frac{1}{2\pi i} \int_C e^{s\tau} ds \frac{1}{(2\pi)^2} \int_0^\infty r dr \int_0^{2\pi} d\theta \frac{-N(ir \sin \theta, ir \cos \theta, s)}{Q(ir \sin \theta, ir \cos \theta, s)} \times$$

$$\exp\left(ir(y \sin \theta + z \cos \theta)\right) \qquad (3.2.2)$$

69

Setting $s = r\sigma$ and using the facts that N is homogeneous of degree 2, while Q is homogeneous of degree 4, gives

$$v_2 = \frac{1}{(2\pi)^2} \int_0^{2\pi} d\theta \int_0^\infty \exp\left(ir(y \sin\theta + z \cos\theta)\right) dr \frac{1}{2\pi i} \times$$

$$\int_C \frac{-N(i \sin\theta, i \cos\theta, \sigma)}{Q(i \sin\theta, i \cos\theta, \sigma)} e^{\sigma r \tau} d\sigma. \tag{3.2.3}$$

Since $Q(i \sin\theta, i \cos\theta, \sigma) = Q(\sin\theta, \cos\theta, -i\sigma)$, it follows from the discussion of Section 1.5 that under the conditions of strict hyperbolicity (which for the geometry of the present problem, applies to all the 13 crystals of Table 1, Section 1.3), there will be four distinct, pure imaginary σ roots of

$$Q(i \sin\theta, i \cos\theta, \sigma) = 0. \tag{3.2.4}$$

Note that this distinctness property of the σ-roots of (3.2.4) holds uniformly in θ. Call these roots, which are simple pole singularities of the Laplace integrand, $\sigma_j = iv_j(\theta)$ where $j = 1, \ldots, 4$. The Laplace inversion is now immediate and gives, for $\tau > 0$,

$$v_2 = \frac{1}{(2\pi)^2} \int_0^{2\pi} d\theta \sum_{j=1}^{4} \frac{-N(i \sin\theta, i \cos\theta, iv_j)}{Q_\sigma(i \sin\theta, i \cos\theta, iv_j)} \times$$

$$\int_0^\infty \exp\left(ir(y \sin\theta + z \cos\theta + v_j\tau)\right) dr. \tag{3.2.5}$$

For situations in which the Q operator is hyperbolic, but not strictly hyperbolic, double σ-roots of (3.2.4) will occur at isolated values of θ (except for case 10 of Section 1.5, which is easily handled separately). Such points pose no real difficulty to the analysis of this section since the elasticity parameters α, β and γ can be perturbed away from a double point situation and then a proper limit taken on the perturbed solution to the problem.

The geometric relationship of the v-roots of

$$Q(\sin\theta, \cos\theta, v) = 0 \tag{3.2.6}$$

to the normal curve (2.2.3) $F(p, q) = 0$ should not be overlooked. In fact for $p = R \cos\theta$ and $q = R \sin\theta$, then a v-root of (3.2.6) is just the reciprocal of the distance of a point on the normal curve from the origin, the same value of θ being used in both instances. This fact is the reason some authors refer to the normal curve as the slowness curve. Thus it is now easy to recognize (from the discussion of Section 2.5) conditions which result in multiple roots of (3.2.4).

Returning now to the evaluation of the double integral (3.2.5), it is clear that the second integral therein does not exist in the classical sense. This integral is of the form

$$I(\lambda) = \int_0^\infty e^{ir\lambda} dr \quad \text{with} \quad \text{Im }\lambda = 0. \tag{3.2.7}$$

Such integrals will be interpreted as

$$I(\lambda) = \lim_{\epsilon \to 0} \int_0^\infty e^{ir\lambda - \epsilon r} dr \qquad (\epsilon > 0). \tag{3.2.8}$$

Then

$$I(\lambda) = \lim_{\epsilon \to 0} \frac{1}{\epsilon - i\lambda} = \lim_{\epsilon \to 0} \frac{\epsilon + i\lambda}{\lambda^2 + \epsilon^2} = \frac{i}{\lambda} + \lim_{\epsilon \to 0} \frac{\epsilon}{\lambda^2 + \epsilon^2} = \frac{i}{\lambda} + \pi\delta(\lambda), \tag{3.2.9}$$

having used a well known delta-convergent sequence definition of the Dirac delta [8]. This interpretation of $I(\lambda)$ gives a result, namely on the right-hand side of equation (3.2.9) which is in agreement with the distributional interpretation of such integrals as given in Gel'fand and Shilov [8].

In view of the above remarks, the second integral in equation (3.2.5) is now given by

$$v_2 = \lim_{\epsilon \to 0} \frac{1}{(2\pi)^2} \int_0^{2\pi} d\theta \sum_{j=1}^4 \frac{-N(i \sin \theta, i \cos \theta, iv_j)}{Q_\sigma(i \sin \theta, i \cos \theta, iv_j)} \times$$

$$\frac{1}{\epsilon - i(y \sin \theta + z \cos \theta + v_j \tau)} \tag{3.2.10}$$

But since $iv_j = \sigma_j$, the integrand of (3.2.10) becomes

$$\sum_{j=1}^4 \frac{-N(i \sin \theta, i \cos \theta, \sigma_j)}{Q_\sigma(i \sin \theta, i \cos \theta, \sigma_j)} \frac{1}{\tau[k(\theta) - \sigma_j]} = \frac{-N(i \sin \theta, i \cos \theta, k)}{\tau Q(i \sin \theta, i \cos \theta, k)}, \tag{3.2.11}$$

where

$$k(\theta) = \frac{y \sin \theta + z \cos \theta + i\epsilon}{i\tau}.$$

Thus the integrand of (3.2.10) is simply the partial fraction expansion of the right-hand side of equation (3.2.11). This is one of the key steps in the inversion process and cleverly avoids the individual evaluation of four (complicated) integrals in (3.2.5) in favor of the much simpler integral

$$v_2 = \lim_{\epsilon \to 0} \frac{1}{4\pi^2\tau} \int_0^{2\pi} \frac{-N(i \sin \theta, i \cos \theta, k)}{Q(i \sin \theta, i \cos \theta, k)} d\theta. \tag{3.2.12}$$

Since the integrand of equation (3.2.12) is periodic in θ (with period 2π) the integration limits $(0, 2\pi)$ may be replaced by $(\pi, 3\pi)$. Then making the substitution $\theta = \phi + \pi$, it is easily shown that the new integral so obtained is the same as (3.2.12) but with ϵ replaced by $-\epsilon$ in the integrand. Thus, using the periodicity of the integrand,

$$v_2 = \lim_{\epsilon \to 0} \frac{1}{4\pi^2\tau} \int_{-\pi}^{\pi} \frac{-N(i \sin \theta, i \cos \theta, m)}{Q(i \sin \theta, i \cos \theta, m)} d\theta, \tag{3.2.13}$$

where

$$m(\theta) = \frac{y \sin \theta + z \cos \theta \pm i\epsilon}{i\tau}.$$

The integrand of (3.2.13) can be simplified further still by using the homogeneity property of N and Q and noting that both of these functions are even in all their members. Hence,

$$v_2 = \lim_{\epsilon \to 0} \frac{1}{4\pi^2 \tau} \int_{-\pi}^{\pi} \frac{N(\sin \theta, \cos \theta, g_\epsilon)}{Q(\sin \theta, \cos \theta, g_\epsilon)} d\theta, \tag{3.2.14}$$

where $g_\epsilon(\theta) = \bar{y} \sin \theta + \bar{z} \cos \theta \pm i\epsilon$, having replaced ϵ/τ by ϵ. Here the (dimensionless) variables $\bar{y}$ and $\bar{z}$ are defined by

$$\bar{y} = y/\tau \quad \text{and} \quad \bar{z} = z/\tau, \tag{3.2.15}$$

and it is worth noting that the v_2 displacement has the functional form

$$v_2(x, y, \tau) = \frac{1}{\tau} f(\bar{y}, \bar{z}).$$

The non-vanishing of ϵ in the integrand of (3.2.14) insures that the integrand remains finite on the θ-integration path. Quoting Burridge [31], another way of saying that is that, '... the integral [for v_2] regarded as a contour integral in the θ-plane must pass around each pole on the contour but not through it in order to avoid divergent integrals. The same effect is accomplished by integrating with respect to θ along a line L_δ parallel to the real θ-axis from $-\pi - i\delta$ to $\pi - i\delta$, and then allowing δ to tend to zero.' Also, since the integral (3.2.14) is independent of the sign of ϵ, the integral must be real (as is obvious physically). Hence there is no loss in generality in working with the real part (Re) of this integral. Thus, upon combining these results, the expression for v_2 becomes

$$v_2 = \lim_{\delta \to 0} \frac{1}{4\pi^2 \tau} \mathrm{Re} \int_{-\pi - i\delta}^{\pi - i\delta} \frac{N(\sin \theta, \cos \theta, \bar{y} \sin \theta + \bar{z} \cos \theta)}{Q(\sin \theta, \cos \theta, \bar{y} \sin \theta + \bar{z} \cos \theta)} d\theta, \tag{3.2.16}$$

where (say) $\delta > 0$. The integral can be split into the form

$$\int_{-\pi - i\delta}^{\pi - i\delta} = \int_{-\pi - i\delta}^{-i\delta} + \int_{-i\delta}^{\pi - i\delta}.$$

In the first integral set $\theta = \phi - \pi$. The first integral then becomes identical with the second, so that

$$v_2 = \lim_{\delta \to 0} \frac{1}{2\pi^2 \tau} \mathrm{Re} \int_{-i\delta}^{\pi - i\delta} \frac{N(\sin \theta, \cos \theta, \bar{y} \sin \theta + \bar{z} \cos \theta)}{Q(\sin \theta, \cos \theta, \bar{y} \sin \theta + \bar{z} \cos \theta)} d\theta. \tag{3.2.17}$$

Consider next the mapping

$$p = \cot \theta \tag{3.2.18}$$

from the complex θ-plane to the complex p-plane. Under this mapping the line segment l_δ in the θ-plane (directed from $-i\delta$ to $\pi - i\delta$) maps into the (large) circle $\bar{C}_\delta$:

$$p_1^2 + [p_2 - \tfrac{1}{2}(\coth \delta + \tanh \delta)]^2 = (\coth \delta - \tanh \delta)^2/4$$

in the p $(= p_1 + ip_2)$ plane, which is traversed in the clockwise direction. Hence, reversing the direction of $\bar{C}_\delta$ and calling the resulting path C_δ (counterclockwise loop) gives

$$v_2 = -\lim_{\delta \to 0} \frac{1}{2\pi^2 \tau} \, \text{Re} \int_{C_\delta} \frac{\sin^2 \theta N(1, p, \bar{y} + \bar{z}p)}{\sin^4 \theta Q(1, p, \bar{y} + \bar{z}p)} \, d\theta.$$

But $dp = d \cot \theta = -d\theta/\sin^2 \theta$, so

$$v_2 = \lim_{\delta \to 0} \frac{1}{2\pi^2 \tau} \, \text{Re} \int_{C_\delta} \frac{N(1, p, \bar{y} + \bar{z}p)}{Q(1, p, \bar{y} + \bar{z}p)} \, dp. \tag{3.2.19}$$

Now as $\delta \to 0$ the circle C_δ encloses, in a positive sense, the upper half p-plane (but definitely excluding the real axis), so that

$$v_2(y, z, \tau) = \frac{1}{\pi \tau} \, \text{Im} \sum_p \left\{ \text{residues of } \frac{-N(1, p, \bar{y} + \bar{z}p)}{Q(1, p, \bar{y} + \bar{z}p)} \text{ in the} \right.$$

$$\left. \text{upper half } p\text{-plane} \right\}. \tag{3.2.20}$$

In a similar way

$$v_3(y, z, \tau) = \frac{1}{\pi \tau} \, \text{Im} \sum_p \left\{ \text{residues of } \frac{\kappa p}{Q(1, p, \bar{y} + \bar{z}p)} \text{ in the} \right.$$

$$\left. \text{upper half } p\text{-plane} \right\}, \tag{3.2.21}$$

$$w_2(y, z, \tau) = v_3(y, z, \tau), \tag{3.2.22}$$

and

$$w_3(y, z, \tau) = \frac{1}{\pi \tau} \, \text{Im} \sum_p \left\{ \text{residues of } \frac{-K(1, p, \bar{y} + \bar{z}p)}{Q(1, p, \bar{y} + \bar{z}p)} \text{ in the} \right.$$

$$\left. \text{upper half } p\text{-plane} \right\}. \tag{3.2.23}$$

From equations (1.6.3)–(1.6.7) it is apparent that the transformation T: $(y, z) \to (z, y)$ together with the parameter switch $(\alpha, \beta) \to (\beta, \alpha)$ leaves the Q and $\partial^2/\partial y \partial z$ operators invariant, while $N \to K$ and $K \to N$. This implies that the displacements $v_2, \ldots, w_3$ may be expressed in terms of the two functions ϕ and χ by

$$v_2(y, z, \tau) \equiv \phi(y, z, \tau | \alpha, \beta, \gamma), \qquad v_3(y, z, \tau) \equiv \chi(y, z, \tau | \alpha, \beta, \gamma),$$

$$w_2(y, z, \tau) = \chi(y, z, \tau | \alpha, \beta, \gamma) \quad \text{and} \quad w_3(y, z, \tau) = \phi(z, y, \tau | \beta, \alpha, \gamma). \tag{3.2.24}$$

Also ϕ is an even function of y and z while χ is odd in y and z. These remarks considerably simplify any numerical study of the displacement functions.

Equations (3.2.20)–(3.2.23) constitute a closed form expression for the 2-dimensional Green's tensor for an unbounded transversely isotropic solid, in which the z-axis is the material symmetry axis. An analysis of these equations will be given below in the next few sections.

Primary references for this section are Burridge [31], and Payton [33, 34]. Budaev [35], using an integration approach involving the complications of branch cuts and Riemann surfaces has also studied the plane strain response of anisotropic materials.

3. The Existence of Lacunas

The real p zeros of the quartic $Q(1, p, \bar{y} + \bar{z}p)$ lie on the real p-plane axis and hence contribute nothing to the residue expressions (3.2.20)–(3.2.23). They do, however, furnish valuable information about the transient displacements, since if there are four such zeros, the displacements will all vanish at the corresponding values of $\bar{y}$ and $\bar{z}$.

First note that the vanishing of $(\bar{y} + \bar{z}p)$, for fixed $\bar{y}$ and $\bar{z}$ cannot lead to a real p zero of Q, since it would simultaneously require that

$$Q(1, p, 0) = \alpha p^4 + \gamma p^2 + \beta = 0. \tag{3.3.1}$$

The p roots of (3.3.1) are given by

$$p^2 = \frac{-\gamma \pm \sqrt{\gamma^2 - 4\alpha\beta}}{2\alpha}, \tag{3.3.2}$$

which are clearly complex for $-2\sqrt{\alpha\beta} < \gamma \leqslant 2\sqrt{\alpha\beta}$. While if $\gamma > 2\sqrt{\alpha\beta}$, then $(\gamma^2 - 4\alpha\beta) > 0$ with the result that $p^2 < 0$ and again there is no real p root of (3.3.1).

Thus from

$$Q(1, p, \bar{y} + \bar{z}p) = (\bar{y} + \bar{z}p)^4 Q(s, r, 1) = (\bar{y} + \bar{z}p)^4 Q(s, r, -1), \tag{3.3.3}$$

where

$$r = \frac{p}{\bar{y} + \bar{z}p} \quad \text{and} \quad s = \frac{1}{\bar{y} + \bar{z}p} \tag{3.3.4}$$

the real p zeros of Q must satisfy

$$Q(s, r, -1) \equiv F(r, s) = 0, \tag{3.3.5}$$

where, from equation (2.2.2), $F(r, s) = 0$ describes the normal curve in the (r, s) plane associated with the elastic parameters α, β and γ. Equations (3.3.4) may be thought of as the p-parametric equation of a curve in the (r, s) plane. At each point where this curve intersects the normal curve, a real p zero of Q is found. Eliminating p from equations (3.3.4) gives

$$\bar{z}r + \bar{y}s = 1$$

which is a straight line. Thus at any point, say (r_0, s_0) where the line $\bar{z}r + \bar{y}s = 1$ intersects the normal curve, the corresponding p value, namely $p = r_0/s_0$, will be a real zero of Q, i.e. $Q(1, p_0, \bar{y} + \bar{z}p_0) = 0$.

Of special importance are the values of $\bar{y}$ and $\bar{z}$ (and α, β, γ) for which $Q(1, p, \bar{y} + \bar{z}p)$ has four real zeros. This can come about in two ways. First if the line $L_1 : \bar{z}r + \bar{y}s = 1$

(see figure 52) intersects both branches of the normal curve (from the condition of strict hyperbolicity, the normal curve consists of two ovals, one inside the other), then there will be four real p zeros. This corresponds physically to values of (y, z, τ) ahead of the leading wavefront where (due to zero initial conditions) all the displacements are zero prior to the arrival of the leading front.

Secondly, if the line $L_2: \bar{z}r + \bar{y}s = 1$ intersects the outer branch N_+ of the normal curve shown in figure 52 at four points, then again there will be four real p zeros. This can only happen for certain (α, β, γ) combinations which lead to concave portions of N_+, since otherwise the normal curve will be convex. It follows from the discussion of sections 2.7 and 2.8 (and in particular equations (2.8.8)–(2.8.9)) that a concave portion of N_+ maps into a cuspidal triangle on W_+. Hence the corresponding $(\bar{y}, \bar{z})$ values associated with L_2 will lie in the interior of this triangular domain, behind the leading branch of the wavefront W_-. Following Petrowski [32], such domains are called *Lacunas* (or gaps). Thus lacunas are zones behind the leading wavefront where Green's tensor is identically zero. The existence of a (spherical) lacuna for a 3-dimensional *isotropic* solid is well known, but that lacunas could exist in 2-dimensional

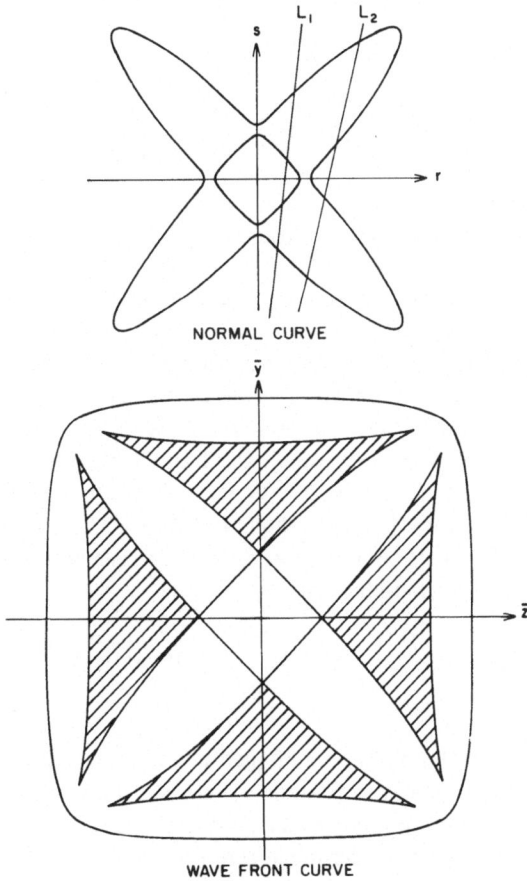

Figure 52. Class IV Normal and Wave Curves. The cross-hatched zones are Lacunas.

75

elastic wave problems is a more recent revelation apparently first noticed by Kraut [36]. Since the lacunas occur for certain values of y/τ and z/τ, then in the plane of the spatial coordinates (y, z) *the lacunas will move with time.*

From the discussion of the normal curve given in sections 2.3 and 2.4, it is now clear that the existence of inflection points on N_+ is a sufficient condition for the existence of lacunas. Thus *wavefronts of classes II–V will give rise to lacunas* for the problem posed in section 3.1. From table 7, section 2.10, it is observed that the hexagonal crystals Apatite, Beryl, Cadmium, Cobalt, Ice, Rhenium and Zinc all give rise to lacunas.

4. Examination of the Complex p-Roots of $Q(1, p, \bar{y}+\bar{z}p)=0$ When the Observation Point is on a Coordinate Axis

In order to perform the residue calculation indicated in equations (3.2.19)–(3.2.23), it is necessary to locate the upper half p-plane zeros of the fourth degree polynomial $Q(1, p, \bar{y} + \bar{z}p)$. From the formula for the 2-dimensional Q operator (1.6.5), it follows that

$$Q(1, p, \bar{y} + \bar{z}p) = [\alpha - (\alpha + 1)\bar{z}^2 + \bar{z}^4]p^4 + [-2(\alpha + 1)\bar{y}\bar{z} + 4\bar{y}\bar{z}^3]p^3$$
$$+ [\gamma - (\alpha + 1)\bar{y}^2 - (\beta + 1)\bar{z}^2 + 6\bar{y}^2\bar{z}^2]p^2$$
$$+ [-2(\beta + 1)\bar{y}\bar{z} + 4\bar{y}^3\bar{z}]p + [\beta - (\beta + 1)\bar{y}^2 + \bar{y}^4]. \quad (3.4.1)$$

Since Q is a quartic in p, in principle equations (3.2.19)–(3.2.23) furnish the means to obtain a closed form solution for the four components of Green's tensor. However, because of the complicated nature of the roots for a (full) quartic equation, this will not be attempted here. There are, however, two interesting cases in which the quartic (3.4.1) reduces to a bi-quadratic, namely $\bar{y} = 0$ and $\bar{z} = 0$. Since the displacements for these two cases are closely related by equations (3.2.24) and the remarks which immediately follow them, results will only be stated below for displacements on the z-axis ($\bar{y} = 0$). Furthermore, since the primary application to be treated here is to the 13 hexagonal crystals of table 7 for which $\alpha > 1$ and $\beta > 1$, only these parameters will be analyzed in detail. It should be mentioned that there are physically interesting plane strain problems, for which equations (3.2.19)–(3.2.23) remain valid and for which $\alpha < 1$ and $\beta < 1$. This could occur, e.g. for crystals with tetragonal symmetry. When $\bar{y} = 0$, equation (3.4.1) reduces to

$$Q(1, p, \bar{z}p) = (\alpha - \bar{z}^2)(1 - \bar{z}^2)p^4 + [\gamma - (\beta + 1)\bar{z}^2]p^2$$
$$+ \beta \equiv f(p). \quad (3.4.2)$$

The expression (3.4.2) for $f(p)$ is a bi-quadratic in p, so that the p-roots may easily be found by using the quadratic formula. There are, however, a number of special cases which depend upon the relative values of the parameter α, β and γ as well as the (moving) observation point $\bar{z}$. Under the assumption

$$\alpha > 1 \quad \text{and} \quad \beta > 1, \quad (3.4.3)$$

and insofar as they pertain to the hexagonal crystals listed in table 7, the upper half p-plane zeros of $f(p)$ will now be catalogued as a function of $|\bar{z}|$ for various ranges of the parameter γ.

$$-2\sqrt{\alpha\beta}<\gamma<(1+\alpha\beta)$$

$\quad |\bar{z}| \geqslant \sqrt{\alpha}$ \qquad no complex roots, \hfill (3.4.4)

$$1<|\bar{z}|<\sqrt{\alpha} \qquad p=p_1 \equiv i\left[\frac{-\{\gamma-(\beta+1)\bar{z}^2\}+D^{1/2}}{-2(\alpha-\bar{z}^2)(1-\bar{z}^2)}\right]^{1/2}, \tag{3.4.5}$$

$$(\beta+1)<\gamma<(1+\alpha\beta)$$

$$|\bar{z}|=1 \qquad p=p_0 \equiv i\left[\frac{\beta}{\gamma-(\beta+1)}\right]^{1/2}, \tag{3.4.6}$$

$$-2\sqrt{\alpha\beta}<\gamma\leqslant(\beta+1)$$

$\quad |\bar{z}|=1$ \qquad no complex roots, \hfill (3.4.7)

$$(\alpha+\beta)<\gamma<(1+\alpha\beta)$$

$\quad 0\leqslant|\bar{z}|<1 \qquad p=p_1 \quad$ and

$$p=p_2 \equiv i\left[\frac{\{\gamma-(\beta+1)\bar{z}^2\}+D^{1/2}}{2(\alpha-\bar{z}^2)(1-\bar{z}^2)}\right]^{1/2}, \tag{3.4.8}$$

$$(\beta+1)<\gamma<(\alpha+\beta)\ and\ (\gamma^2-4\alpha\beta)<0$$

$\quad \bar{z}_1<|\bar{z}|<1 \qquad p=p_1 \quad$ and $\quad p=p_2, \hfill (3.4.9)$

$$|\bar{z}|=\bar{z}_1 \qquad p=p_4 \equiv i\left[\frac{\beta}{(\alpha-\bar{z}_1^2)(1-\bar{z}_1^2)}\right]^{1/4} \qquad \text{(double root)} \quad (3.4.10)$$

$\quad 0\leqslant|\bar{z}|<\bar{z}_1 \qquad p=p_+ \quad$ and $\quad p=p_- \hfill (3.4.11)$

where

$$P_{\pm} = \pm\left[\frac{2\{\beta(\alpha-\bar{z}^2)(1-\bar{z}^2)\}^{1/2}-\{\gamma-(\beta+1)\bar{z}^2\}}{4(\alpha-\bar{z}^2)(1-\bar{z}^2)}\right]^{1/2}$$

$$+i\left[\frac{2\{\beta(\alpha-\bar{z}^2)(1-\bar{z}^2)\}^{1/2}+\{\gamma-(\beta+1)\bar{z}^2\}}{4(\alpha-\bar{z}^2)(1-\bar{z}^2)}\right]^{1/2}. \tag{3.4.12}$$

$$\gamma<(\beta+1)\ and\ (\gamma^2-4\alpha\beta)<0$$

$$\bar{z}_1 \leqslant |\bar{z}| < 1 \qquad \text{no complex roots,} \tag{3.4.13}$$

$$0 \leqslant |\bar{z}| < \bar{z}_1 \qquad p = p_+ \quad \text{and} \quad p = p_-. \tag{3.4.14}$$

In the above expressions, D is the discriminant of the bi-quadratic $f(p)$ and is given by

$$D(\bar{z}) = \{\gamma - (\beta + 1)\bar{z}^2\}^2 - 4\beta(\alpha - \bar{z}^2)(1 - \bar{z}^2). \tag{3.4.15}$$

Also

$$D(\bar{z}_1) = 0, \tag{3.4.16}$$

where

$$\bar{z}_1 = [\gamma(\beta + 1) - 2\beta(\alpha + 1) + 2\{\beta(1 + \alpha\beta - \gamma)(\alpha + \beta - \gamma)\}^{1/2}]^{1/2}/(\beta - 1). \tag{3.4.17}$$

For the γ-range $\gamma < (\beta + 1)$ and $(\gamma^2 - 4\alpha\beta) < 0$, $\bar{z}_1$ has an interesting geometric interpretation. At the points $z = \pm \bar{z}_1 \tau$ on the z-axis, the wave front (W_+) crosses itself. From the discussion of section 2.8, this implies that the normal curve $Q(q, p, -1) = 0$ has a bitangent at $p^{-1} = \bar{z}_1$.

The above analysis of the complex roots of $f(p) = 0$, while not exhaustive, covers all the crystals listed in table 7. The results can be easily extended to other situations for which the inequalities (3.4.3) do not apply. One notable omission is the case $\gamma = (\alpha + \beta)$, which will be treated below (section 3.10) for a 3-dimensional solid. This case includes the isotropic solid.

5. Displacement Components Along the Symmetry Axis for Some Hexagonal Materials

With the information about the poles of $Q^{-1}(1, p, \bar{z}p)$ from section 3.4, it is now an easy matter to perform the residue calculations called for in equations (3.2.19)–(3.2.22). The displacement components along the $\bar{z}$-axis may be expressed in terms of the six functions $F_1, \ldots, G_3$

$$F_1(\bar{z}) = \left[\frac{1}{4\beta} - \frac{2\beta(\alpha - \bar{z}^2) - \{\gamma - (\beta + 1)\bar{z}^2\}}{4\beta D^{1/2}}\right]\left[\frac{-\{\gamma - (\beta + 1)\bar{z}^2\} + D^{1/2}}{-2(\alpha - \bar{z}^2)(1 - \bar{z}^2)}\right]^{1/2}, \tag{3.5.1}$$

$$F_2(\bar{z}) = \left[\frac{1}{4\beta} + \frac{2\beta(\alpha - \bar{z}^2) - \{\gamma - (\beta + 1)\bar{z}^2\}}{4\beta D^{1/2}}\right]\left[\frac{\{\gamma - (\beta + 1)\bar{z}^2\} + D^{1/2}}{2(\alpha - \bar{z}^2)(1 - \bar{z}^2)}\right]^{1/2}, \tag{3.5.2}$$

$$F_3(\bar{z}) = \frac{1}{2}\left[\frac{1}{\sqrt{\beta}} + \sqrt{\frac{\alpha - \bar{z}^2}{1 - \bar{z}^2}}\right][\{\gamma - (\beta + 1)\bar{z}^2\} + 2\{\beta(\alpha - \bar{z}^2)(1 - \bar{z}^2)\}^{1/2}]^{-1/2} \tag{3.5.3}$$

$$G_1(\bar{z}) = \left[\frac{1}{4} - \frac{2(1 - \bar{z}^2) - \{\gamma - (\beta + 1)\bar{z}^2\}}{4D^{1/2}}\right]\left[\frac{-\{\gamma - (\beta + 1)\bar{z}^2\} + D^{1/2}}{-2(\alpha - \bar{z}^2)(1 - \bar{z}^2)}\right]^{1/2}, \tag{3.5.4}$$

$$G_2(\bar{z}) = \left[\frac{1}{4} + \frac{2(1-\bar{z}^2) - \{\gamma - (\beta+1)\bar{z}^2\}}{4D^{1/2}} \right] \left[\frac{\{\gamma - (\beta+1)\bar{z}^2\} + D^{1/2}}{2(\alpha-\bar{z}^2)(1-\bar{z}^2)} \right]^{1/2},$$

(3.5.5)

and

$$G_3(\bar{z}) = \frac{1}{2} \left[\sqrt{\beta} + \sqrt{\frac{1-\bar{z}^2}{\alpha-\bar{z}^2}} \right] [\{\gamma - (\beta+1)\bar{z}^2\} + 2\{\beta(\alpha-\bar{z}^2)(1-\bar{z}^2)\}^{1/2}]^{-1/2}$$

(3.5.6)

where $D(\bar{z})$ has previously been defined in equation (3.4.15). From the remarks of section 3.2 (i.e., the displacements $w_2 = v_3$ are odd functions of $\bar{y}$) it follows that

$$v_3(0, z, \tau) = w_2(0, z, \tau) \equiv 0,$$

(3.5.7)

so that it is only necessary to list the displacements $v_2(0, z, \tau)$ and $w_3(0, z, \tau)$. Calling

$$\pi\tau v_2(0, z, \tau) \equiv V(\bar{z}) \quad \text{and} \quad \pi\tau w_3(0, z, \tau) \equiv W(\bar{z}),$$

(3.5.8)

then, depending upon the appropriate γ-range, the nonvanishing displacements along the $\bar{z}$-axis are given by

(α + β) < γ < (1 + αβ)

$$
\begin{array}{lll}
|\bar{z}| \geqslant \sqrt{\alpha} & V = 0, & W = 0 \\
1 \leqslant |\bar{z}| < \sqrt{\alpha} & V = F_1, & W = G_1 \\
0 \leqslant |\bar{z}| < 1 & V = F_1 + F_2, & W = G_1 + G_2
\end{array}
$$

(3.5.9)

(β + 1) < γ < (α + β) and (γ² − 4αβ) < 0

$$
\begin{array}{lll}
|\bar{z}| \geqslant \sqrt{\alpha} & V = 0, & W = 0 \\
1 \leqslant |\bar{z}| < \sqrt{\alpha} & V = F_1, & W = G_1 \\
\bar{z}_1 \leqslant |\bar{z}| \leqslant 1 & V = F_1 + F_2, & W = G_1 + G_2 \\
0 \leqslant |\bar{z}| \leqslant \bar{z}_1 & V = F_3, & W = G_3
\end{array}
$$

(3.5.10)

γ < (β + 1) and (γ² − 4αβ) < 0

$$
\begin{array}{lll}
|\bar{z}| \geqslant \sqrt{\alpha} & V = 0, & W = 0 \\
1 < |\bar{z}| < \sqrt{\alpha} & V = F_1, & W = G_1 \\
\bar{z}_1 \leqslant |\bar{z}| \leqslant 1 & V = 0 \text{ (lacuna)}, & W = 0 \text{ (lacuna)} \\
\bar{z}_1 < |\bar{z}| \leqslant 0 & V = F_3, & W = G_3.
\end{array}
$$

(3.5.11)

In Table 11, the values of α, β and γ from table 7 are used to classify the 13 crystals according to the γ-ranges indicated by equations (3.5.9)–(3.5.11).

Figure 53 shows the variation of the displacements $\pi\tau v_2(0, z, \tau)$ $(= V(\bar{z}))$ and

Table 11.   Hexagonal Crystal Classification Appropriate to Symmetry Axis Displacements.

Crystal	Remark	
Apatite	(3)	(1) $(\alpha + \beta) < \gamma < (1 + \alpha\beta)$.
Beryllium	(2)	(2) $(\beta + 1) < \gamma < (\alpha + \beta)$ and $(\gamma^2 - 4\alpha\beta) < 0$.
Beryl	(1)	(3) $\gamma < (\beta + 1)$ and $(\gamma^2 - 4\alpha\beta) < 0$ also $\beta > \alpha$.
Cadmium	(3)	
Cobalt	(1)	
Ice	(1)	
Hafnium	(1)	
Magnesium	(1)	
Rhenium	(1)	
Titanium	(1)	
Thallium	(1)	
Yttrium	(1)	
Zinc	(3)	

$\pi\tau w_3(0, z, \tau) (= W(\bar{z}))$ as a function of $\bar{z}$, for Beryl, Beryllium and Zinc. These crystals are representative of the three γ-ranges and displacement expressions given by equations (3.5.9)–(3.5.11).

When $(\alpha + \beta) < \gamma < (1 + \alpha\beta)$ the $\bar{z}$-axis is intersected by two wave fronts at $|\bar{z}| = \sqrt{\alpha}$ and $|\bar{z}| = 1$ (e.g., Beryl in figure 53, see also figure 21 for the Beryl wavefront). $W(\bar{z})$ has an inverse square root singularity as $\bar{z}$ approaches the leading front, while $V(\bar{z})$ tends to zero but with a (positive) infinite slope. $V(\bar{z})$ has an inverse square root

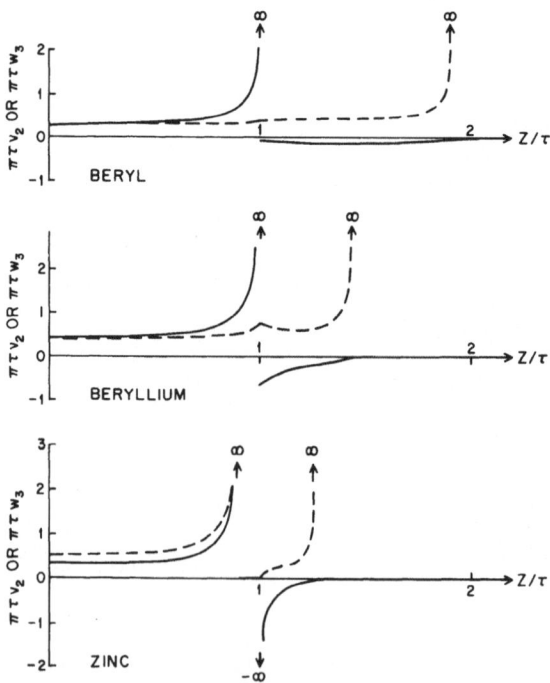

Figure 53.   Displacement Components v_2 (———) and w_3 (— — —) Versus z/τ Along the Line $y = 0$.

singularity at the slower front when approached from below, while it approaches a finite value with a slope whose sign is given by (for $\beta > 4/3$)

$$\text{sgn} \left[5/(3\beta + 1) - (\gamma - \beta - 1)/(\beta\alpha - \beta) \right],\qquad(3.5.12)$$

as $\bar{z}$ approaches 1 from above. $W(\bar{z})$ is continuous at $\bar{z} = 1$ but has a (positive) infinite slope as $\bar{z}$ approaches 1 from below, while the sign of the slope is given by $\text{sgn}\,(\gamma - \bar{\gamma})$, where $\bar{\gamma} = 1 + \beta(\beta + 3\alpha)/(\beta + 3)$, as $\bar{z}$ approaches 1 from above. This slope behaviour for $W(\bar{z})$ at $\bar{z} = 1$ can result in a kink in the displacement curve.

When $(\beta + 1) < \gamma < (\alpha + \beta)$ and $(\gamma^2 - 4\alpha\beta) < 0$, as is the case for Beryllium (see figure 53 and figure 19 for the wavefront), the remarks of the previous paragraph apply. Note in particular the kink in the $W(\bar{z})$ curve at $\bar{z} = 1$ for Beryllium. From the form change of the solution at $\bar{z} = \bar{z}_1$ it might be expected that the displacements or their derivatives (of some order) would be discontinuous at this point. However this is not so, both $V(\bar{z})$ and $W(\bar{z})$ are infinitely differentiable at $\bar{z} = \bar{z}_1$.

Finally when $\gamma < (\beta + 1)$ and $(\gamma^2 - 4\alpha\beta) < 0$ the displacement behaviour at the leading front $|\bar{z}| = \sqrt{\alpha}$ is the same as above (e.g., Zinc in figure 53, see also figure 43 for the Zinc wavefront). As $\bar{z}$ approaches 1 (in this case the leading edge of the lacuna) from above, $V(\bar{z})$ experiences an inverse square root singularity while W tends to zero with a (positive) infinite slope. Both $V(\bar{z})$ and $W(\bar{z})$ have inverse square root singularities when the trailing edge of the lacuna ($\bar{z} = \bar{z}_1 = 0.8871$ for Zinc) is approached from below. Incidentally the latter point is a double point on the wavefront.

The above descriptive remarks regarding the behaviour of the displacement functions $V(\bar{z})$ and $W(\bar{z})$ are not altogether obvious from the expressions (3.5.9)–(3.5.11), but are rather based on expansions of these functions in the neighborhood of the wavefront in question. The discussion of this section closely follows that of Payton [34].

6. The Solution Behaviour Near the Wavefront

Although equations (3.2.20)–(3.2.23) constitute an exact solution to the Green's tensor problem for an unbounded solid, their interpretation in a general situation is made difficult by the necessity of finding the roots of the quartic (3.4.1). In section 3.4 it was pointed out that this quartic simplifies to a bi-quadratic when the displacements are evaluated along either the $\bar{y}$ or $\bar{z}$ coordinate axes, but otherwise the full quartic must be investigated. In the present section the exact solution will be analyzed in the immediate neighborhood of the wavefront. It will be shown that in general (there are exceptions) the displacements become unbounded as W is approached from the left-hand side when the front is traced in the direction of increasing θ, where θ is the parameter along W. At an ordinary point of W (to be defined below) where the displacement is singular, the singularity is of the order $d^{-1/2}$ where d measures distance along the normal vector away from W at the point in question on W. However at the cusp tip of a lacuna, this singularity is of order $d^{-2/3}$ except when the cusp tip is approached along a special direction which allows the lacuna to be penetrated. This will be explained more fully below.

In order to extract information about the displacement singularities at the wavefront (from the theory of weak solutions to linear partial differential equations [18], it is known that there can be no others) a perturbation technique will be employed on the exact solution. This approach may be contrasted with that of Bazer and Yen [37] who base their analysis on certain asymptotic formulae of Ludwig [38]. Also, of particular interest in this connection is the paper of Willis [39].

Perturbation Technique for Determining the Singularities on W

In section 3.3 it was shown that, for given values of $(\bar{z}, \bar{y})$ and the parameters (α, β, γ), the real p-plane roots of

$$Q(1, p, \bar{y} + \bar{z}p) \equiv f(p|\bar{z}, \bar{y}), \tag{3.6.1}$$

can be found from a knowledge of the intersection of the points on the line L: $\bar{z}r + \bar{y}s = 1$ with the normal curve N: $Q(s, r, -1) = 0$ in the (r, s) plane. In particular if (r_0, s_0) is an intersection point of L with N, then $p_0 = r_0/s_0$ is a real zero of f.

Now suppose, e.g., that $\bar{z} = \bar{z}_-(\theta)$ and $\bar{y} = \bar{y}_-(\theta)$ where $\bar{z}_-$ and $\bar{y}_-$ have been previously defined in equations (2.8.8) and (2.8.9) as wavefront coordinates on W_-. Then L will touch N_- (the inner branch of N) at $[r_-(\theta), s_-(\theta)]$ and also cross N_+ at two additional distinct points. The fact that L touches (is tangent to) N_- in this case clearly means that $p = p_0 = \cot \theta$ is a double root of $f(p|\bar{z}_-, \bar{y}_-) = 0$. If the values of $\bar{z}$ and $\bar{y}$ are now slightly perturbed away from $\bar{z}_-$ and $\bar{y}_-$, then this double p root will break up into two real roots (if $\bar{z}$ and $\bar{y}$ lie to the right of W_- as this front is traced in the direction of increasing θ) or into two complex roots (if $\bar{z}$ and $\bar{y}$ lie to the left of W_-). Furthermore this perturbation if sufficiently small, will not alter the character of the two additional p roots determined from the intersection of L with N_+. Thus *the behaviour of the displacement functions, near the point $(\bar{z}_-, \bar{y}_-)$ on W_-, can be found by locating the approximate roots of $f(p|\bar{z}, \bar{y}) = 0$ for small values of $(\bar{z} - \bar{z}_-)^2 + (\bar{y} - \bar{y}_-)^2$.* Also, only those two roots which coalesce to p_0 when $(\bar{z}, \bar{y})$ moves to a point on W_- are needed if attention is directed toward the displacement singularity at $(\bar{z}_-, \bar{y}_-)$ when approached from the left side of W_-. This is so since the real character of the two additional roots is not disturbed by the perturbation and hence these roots will not contribute to the displacement expressions.

Since $p_0 = \cot \theta$ is a double root of $f(p|\bar{z}_-, \bar{y}_-)$ then

$$f(p_0|\bar{z}_-, \bar{y}_-) = f'(p_0|\bar{z}_-, \bar{y}_-) = 0. \tag{3.6.2}$$

Realizing that $(\bar{z} - \bar{z}_-)$, $(\bar{y} - \bar{y}_-)$ and $(p - p_0)$ are small (but not of the same order) a triple Taylor series expansion of $f(p|\bar{z}, \bar{y})$ about $f(p_0|\bar{z}_-, \bar{y}_-)$ gives (retaining only the significant terms)

$$f(p|\bar{z}, \bar{y}) \approx H + f''(p_0|\bar{z}_-, \bar{y}_-)(p - p_0)^2/2, \tag{3.6.3}$$

where

$$H = (\bar{z} - \bar{z}_-)f_{\bar{z}}(p_0|\bar{z}_-, \bar{y}_-) + (\bar{y} - \bar{y}_-)f_{\bar{y}}(p_0|\bar{z}_-, \bar{y}_-). \tag{3.6.4}$$

The perturbed roots of $f(p|\bar{z}, \bar{y}) = 0$ are then given by

$$p = p_0 \pm (-2H/f'')^{1/2}. \tag{3.6.5}$$

plus two additional real roots, which contribute nothing to the residue calculation of the displacement functions. Whence, from (3.2.20),

$$v_2 \approx -\frac{1}{\pi\tau} \operatorname{Im} \frac{N(1,p_0,\bar{y}_- + \bar{z}_- p_0)}{(p - p_0)f''} = -\frac{1}{\pi\tau} \operatorname{Im} \frac{N(1,p_0,\bar{y}_- + \bar{z}_- p_0)}{\pm f''[-2H/f'']^{1/2}}. \quad (3.6.6)$$

A point on $W_\pm$ for which $f''(p_0|\bar{z}_\pm,\bar{y}_\pm) \neq 0$ will be called an *ordinary point* of $W_\pm$. Clearly at an ordinary point of W_- the displacement v_2 will have a singularity like $|H|^{-1/2}$ as $(\bar{z}-\bar{z}_-)^2 + (\bar{y}-\bar{y}_-)^2 \to 0$, provided the associated perturbed pole lies in the upper half p-plane. Obviously to treat the situation in more detail requires a thorough knowledge of the sign of $-2H/f''$ for points on W_-.

With but minor alterations, the above argument also applies to the determination of the displacement singularities at ordinary points on W_+.

Expressions for f'' and H

Exploiting the two relations given in equation (3.6.2) (and similar relations for points on W_+) allows f'' to be written as

$$f''(p_0|\bar{z}_\pm,\bar{y}_\pm) = \frac{M_\pm}{A}\left[2\beta \sin^2\theta + \gamma \cos^2\theta \mp \frac{\cos^2\theta\,(k_1\cos^2\theta - k_2\sin^2\theta)}{\sqrt{B^2-4A}}\right]$$

$$+ 8\cos^2\theta \sin^2\theta\, \frac{[\gamma - (\alpha\beta + 1)]\,[\gamma - (\alpha + \beta)]}{(B^2-4A)}, \quad (3.6.7)$$

where

$$M_\pm(\theta) = [2\alpha\cos^2\theta + \gamma\sin^2\theta - (\alpha + 1)V_\pm^2].$$

In the above, the constants k_1 and k_2 have been defined below equation (2.8.9), $A(\theta)$ and $B(\theta)$ are given in (1.5.26) and (1.5.27) and $V(\theta)$ was defined by equation (2.7.19). While the expression (3.6.7) for f'' shows its explicit dependence on θ, it is somewhat complicated and in particular the sign of f'' is not readily available.

An alternate expression for f'' which relates its sign to geometric properties of the normal curve will now be established. From equations (3.3.3)–(3.3.5)

$$f(p|\bar{z},\bar{y}) = (\bar{y} + \bar{z}p)^4 F(r, s) \quad (3.6.8)$$

where $r(p)$ and $s(p)$ are given by (3.3.4). Then, at a double root of f

$$f''(p_0|\bar{z}_\pm,\bar{y}_\pm) = (\bar{y} + \bar{z}p)^4 \frac{d^2F}{dp^2} \quad (3.6.9)$$

it being understood that here and in what follows, all quantities are evaluated at $p = p_0$, $\bar{z} = \bar{z}_\pm$ and $\bar{y} = \bar{y}_\pm$ after the indicated differentiation. Using the chain rule for differentiation of F gives

$$\frac{d^2F}{dp^2} = F_{rr}r_p^2 + 2F_{rs}r_p s_p + F_{ss}s_p^2. \quad (3.6.10)$$

Taking d/dr and d^2/dr^2 of the implicit equation of N: $F(r, s) = 0$ gives

$$\frac{ds}{dr} = -\frac{F_r}{F_s} \quad \text{and} \quad \frac{d^2 s}{dr^2} = -\frac{F_{rr} F_s^2 - 2F_{rs} F_r F_s + F_{ss} F_r^2}{F_s^3}, \tag{3.6.11}$$

so that equation (3.6.10) may be written

$$\frac{d^2 F}{dp^2} = -F_s \left(\frac{dr}{dp}\right)^2 \frac{d^2 s}{dr^2}, \tag{3.6.12}$$

which is valid at an ordinary point on N. Writing, from section 2.2,

$$V_\pm^{-1}(\theta) = R_\pm(\theta) = \left[\frac{B(\theta) \pm \sqrt{B^2 - 4A}}{2A(\theta)}\right]^{1/2}, \tag{3.6.13}$$

together with the polar coordinate θ-parameterization of $N_\pm$

$$r_\pm(\theta) = R_\pm(\theta) \cos \theta \quad \text{and} \quad s_\pm(\theta) = R_\pm(\theta) \sin \theta, \tag{3.6.14}$$

allows $d^2 s/dr^2$ on N to be written in terms of θ derivatives of V

$$\frac{d^2 s}{dr^2} = -\frac{V^3}{(V \sin \theta + V' \cos \theta)^3} (V + V''). \tag{3.6.15}$$

After some simplification, equation (3.6.9) may be rewritten as

$$f''(p_0 | \bar{z}_\pm, \bar{y}_\pm) = \pm 2V_\pm \sqrt{B^2 - 4A} \, (V_\pm + V_\pm''). \tag{3.6.16}$$

Recall (e.g. Stoker [40]) that the curvature of a plane curve specified parametrically by $z = z(\theta)$ and $y = y(\theta)$ in the (z, y) plane is

$$k = \frac{z'y'' - z''y'}{(z'^2 + y'^2)^{3/2}}.$$

Hence, using equations (3.6.14), (2.8.8) and (2.8.9), normal curve curvature

$$k_N = V^3 \frac{V + V''}{(V^2 + V'^2)^{3/2}}, \tag{3.6.17}$$

and wavefront curvature

$$k_W = \frac{1}{|V + V''|}. \tag{3.6.18}$$

Thus the sign of f'' is directly related to the sign of $(V + V'')$ on N which in turn, from (3.6.17), has the same sign as the curvature of the normal curve N.

From the preceeding remarks it is now clear that for a given ordinary point on (say) W_-, f'' will be positive if $k_{N_-} < 0$ and negative if $k_{N_-} > 0$. Since N_- is convex,

$$(V_- + V_-'') > 0, \tag{3.6.19}$$

but $(V_+ + V_+'')$ will change sign at a (simple) inflection point of N_+ where $(V_+ + V_+'') = 0$.

The expression for H given in (3.6.4) may be simplified with the help of the formulae of this section to

$$H = -\frac{2V_\pm}{\sin^4\theta} (\pm\sqrt{B^2 - 4A})\,\psi_\pm, \tag{3.6.20}$$

where

$$\psi_\pm(\theta) = [(\bar{z} - \bar{z}_\pm)\cos\theta + (\bar{y} - \bar{y}_\pm)\sin\theta]. \tag{3.6.21}$$

Equation (3.6.3) for the approximate form of $f(p|\bar{z}, \bar{y})$ when p is close to a double real zero and $(\bar{z}, \bar{y})$ is close to a point on W_+ or W_- becomes

$$f(p|\bar{z}, \bar{y}) \approx \pm 2V_\pm\sqrt{B^2 - 4A}\left[-\frac{\psi_\pm}{\sin^4\theta} + (V_\pm + V_\pm'')(p - p_0)^2/2\right]. \tag{3.6.22}$$

In passing it should be mentioned that with the help of (3.6.7) and (3.6.16), an explicit expression for $(V_\pm + V_\pm'')$ may be obtained.

By the examination of some typical normal curves, it is easily seen that

$$\psi_{-L}(\theta) < 0 \quad \text{and} \quad \psi_{-R}(\theta) > 0, \tag{3.6.23}$$

where $\psi_{-L}(\theta)$ signifies that $(\bar{z}, \bar{y})$ lies immediately to the left of W_- as W_- is traced in the direction of increasing θ, while $\psi_{-R}(\theta)$ has the same interpretation when $(\bar{z}, \bar{y})$ lies immediately to the right of W_-. Similar results for the sign of $\psi_+(\theta)$ are not so simple since $\psi_{+R}(\theta)$ and $\psi_{+L}(\theta)$ may change signs as W_+ is traced. Fortunately any sign change in ψ_+ is coincident with that of $(V_+ + V_+'')$.

The results of this subsection may now be summarized as follows:

$$f(p|\bar{z}, \bar{y}) \approx \pm 2V_\pm\sqrt{B^2 - 4A}\,(V_\pm + V_\pm'')\left[\frac{(p - p_0)^2}{2} - \frac{\chi_\pm(\theta)}{\sin^4\theta}\right], \tag{3.6.24}$$

where

$$\chi_\pm(\theta) = \frac{(\bar{z} - \bar{z}_\pm)\cos\theta + (\bar{y} - \bar{y}_\pm)\sin\theta}{V_\pm + V_\pm''}. \tag{3.6.25}$$

This expression for $f(p|\bar{z}, \bar{y})$ is approximately true for p near a double real root $(p = p_0 = \cot\theta)$ of $f(p|\bar{z}_\pm, \bar{y}_\pm) = 0$ with $(\bar{z}, \bar{y})$ very close to $(\bar{z}_-, \bar{y}_-)$ (i.e. close to W_-) or $(\bar{z}, \bar{y})$ very close to $(\bar{z}_+, \bar{y}_+)$ (i.e. close to W_+). Furthermore

$$\chi_\pm(\theta) < 0 \quad \text{when} \quad (\bar{z}, \bar{y}) \text{ lies to the } left \text{ of } W_\pm, \tag{3.6.26}$$

and

$$\chi_\pm(\theta) > 0 \quad \text{when} \quad (\bar{z}, \bar{y}) \text{ lies to the } right \text{ of } W_\pm. \tag{3.6.27}$$

Displacement Singularity Formulae at an Ordinary Point of W

The displacement function behaviour near the wavefront will now be expressed by formulae which measure the *jump in displacement* at any point on the waterfront, as the point $(\bar{z}, \bar{y})$ *crosses the front.* This jump will be denoted by

$$[v_2]_\pm = v_2(\bar{z} \text{ and } \bar{y} \text{ immediately to the left of } W_\pm)$$

$$-v_2(\bar{z} = \bar{z}_\pm(\theta), \ \bar{y} = \bar{y}_\pm(\theta)). \qquad (3.6.28)$$

where the second term is to be considered as a limit expression with $(\bar{z}, \bar{y})$ approaching $(\bar{z}_\pm, \bar{y}_\pm)$ from the right-hand side of $W_\pm$ when this curve is traced in the direction of increasing θ (see figure 54). This limit will always exist at an ordinary point of $W_\pm$.

In most instances it would be just as easy to give the dominating part of the displacement expression as the front is approached from the left or from the right. as opposed to the difference between these two terms. However, since the displacement is generally nonsingular on the right side of W, the two additional (unperturbed) p roots of $f = 0$ would be required in order to give the limiting value of the displacement as W_+ is approached from the right. Although the limiting form of the displacement function as W is approached from the right and left would provide some additional information, it would contribute nothing further to our knowledge of the singularities involved at W. For this reason, primary attention is directed herein toward the jump in displacement across W. In this connection it is well to point out the significance of

$$\psi_+(\theta) = (\bar{z} - \bar{z}_+) \cos \theta + (\bar{y} - \bar{y}_+) \sin \theta = 0 \qquad (3.6.29)$$

In the $(\bar{z}, \bar{y})$ plane, this expression represents a line L, which passes through the point $(\bar{z}_+, \bar{y}_+)$ on W_+, and has the slope $-\cot \theta$. Now the tangent vector to W_+ at an ordinary

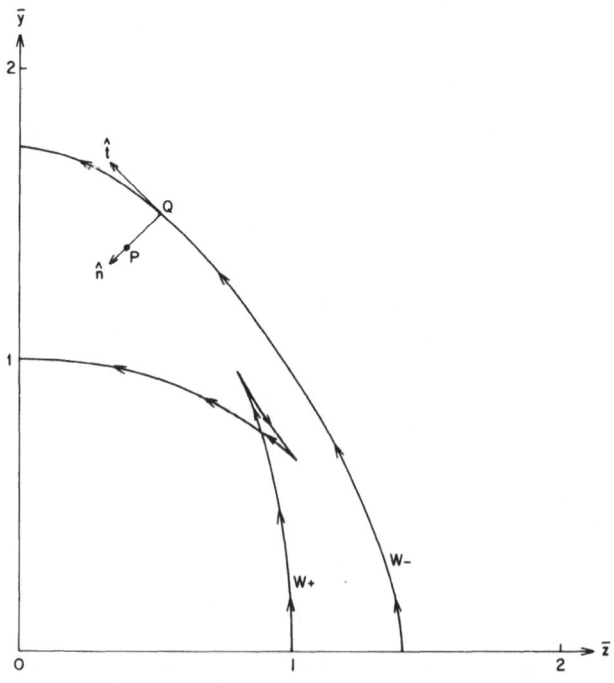

Figure 54.   Direction in Which the Wavefront W is Traced as the Parameter θ increases.

point has the slope

$$\frac{d\bar{y}}{d\bar{z}} = \frac{\bar{y}'_+}{\bar{z}'_+} = \frac{(V_+ + V''_+)\cos\theta}{-(V_+ + V''_+)\sin\theta} = -\cot\theta. \tag{3.6.30}$$

Hence the line L is tangent to W_+ at $(\bar{z}_+, \bar{y}_+)$ and *does not cross* W_+ *at an ordinary point*. Similar remarks apply to the expression $\psi_-(\theta)$.

From equation (3.6.22), the perturbed poles of $f^{-1}(p|\bar{z}, \bar{y})$ are given by

$$\text{LEFT OF } W_\pm \qquad p = p_0 \pm i \frac{\sqrt{2}}{\sin^2\theta} |\chi_\pm|^{1/2}.$$

and the upper half p-plane pole is

$$p = p_0 + i \frac{\sqrt{2}}{\sin^2\theta} |\chi_\pm|^{1/2} \tag{3.6.31}$$

$$\text{RIGHT OF } W_\pm \qquad p = p_0 \pm \frac{\sqrt{2}}{\sin^2\theta} \sqrt{\chi_\pm},$$

hence there are no (perturbed) upper half p-plane poles. From equation (3.2.20) a residue calculation now gives

$$v_2|_{\text{LEFT}} = -\frac{1}{\pi\tau} \operatorname{Im} \frac{N(1, p_0, \bar{y}_\pm + \bar{z}_\pm p_0)}{\pm 2V_\pm \sqrt{B^2 - 4A}(V_\pm + V''_\pm) i \dfrac{\sqrt{2}}{\sin^2\theta}} |\chi_\pm|^{-1/2},$$

$$= \pm \frac{N(\sin\theta, \cos\theta, V_\pm)}{2\sqrt{2\pi\tau}V_\pm \sqrt{B^2 - 4A}(V_\pm + V''_\pm)} |\chi_\pm|^{-1/2}. \tag{3.6.32}$$

Set

$$h_\pm(\theta) = \pm [2\sqrt{2\pi\tau}\sqrt{B^2 - 4A}(V_\pm + V''_\pm)]^{-1}, \tag{3.6.33}$$

then the displacement component jumps near an ordinary point on $W_\pm$ are given by

$$[v_2]_\pm = h_\pm(\theta)N(\sin\theta, \cos\theta, V_\pm)|\chi_\pm(\theta)|^{-1/2} \tag{3.6.34}$$

$$[v_3]_\pm = [w_2]_\pm = -h_\pm(\theta)\kappa \sin\theta \cos\theta |\chi_\pm(\theta)|^{-1/2} \tag{3.6.35}$$

and

$$[w_3]_\pm = h_\pm(\theta)K(\sin\theta, \cos\theta, V_\pm)|\chi_\pm(\theta)|^{-1/2}. \tag{3.6.36}$$

Now if $\mathbf{x}(\theta) = (\bar{z}(\theta), \bar{y}(\theta))$ for $0 \leqslant \theta < 2\pi$ denotes the position vector on W, then it readily follows from (2.8.5) by differentiation that

$$\hat{t}(\theta) = (-\sin\theta, \cos\theta) \tag{3.6.37}$$

is a unit tangent vector to W and

$$\hat{n}(\theta) = (-\cos\theta, -\sin\theta) \tag{3.6.38}$$

is a unit normal vector to W. (See figure 54 where, however, $\hat{t}$ and $\hat{n}$ are not drawn to scale.) Let

$$d = \sqrt{(\bar{z} - \bar{z}_-(\theta))^2 + (\bar{y} - \bar{y}_-(\theta))^2} \tag{3.6.39}$$

be the distance from the point $P(\bar{z}, \bar{y})$ (which lies to the immediate left of W_-) to the point $Q(\bar{z}_-, \bar{y}_-)$ on W_-, so that

$$\mathbf{QP} = (\bar{z} - \bar{z}_-, \bar{y} - \bar{y}_-) \quad \text{and} \quad d = |\mathbf{QP}|. \tag{3.6.40}$$

But also

$$\mathbf{QP} = d\hat{n}, \tag{3.6.41}$$

hence

$$d = |\mathbf{PQ} \cdot \hat{n}| = |(\bar{z} - \bar{z}_-)\cos\theta + (\bar{y} - \bar{y}_-)\sin\theta|. \tag{3.6.42}$$

A similar formula holds for points near W_+.

Since from (3.6.35) χ is proportional to d, it follows from the jump expressions (3.6.34)–(3.6.36) that the displacement jump across the wavefront at an ordinary point on $W_\pm$ is singular as $d^{-1/2}$, d measuring the distance from $(\bar{z}_\pm, \bar{y}_\pm)$ on the left side of $W_\pm$.

Further analysis shows that at points on $W_\pm$ where

$$\sin\theta\cos\theta = 0 \tag{3.6.43}$$

the jump in v_3 (and w_2) is zero. This is not unexpected since these points occur on the $\bar{z}$ and $\bar{y}$-axes where the wavefront is pierced. But since v_3 is an odd function of $\bar{z}$ and $\bar{y}$, $v_3 \equiv 0$ along these axes. In a similar way the expressions $N(\sin\theta, \cos\theta, V_\pm)$ may vanish for certain values of θ and thus smooth out the singularity in v_2 at such points. It is easily shown that a necessary condition for N to vanish is

$$[\gamma - (1 + \alpha\beta)]\sin^2\theta\cos^2\theta = 0. \tag{3.6.44}$$

This result is established by finding the θ-values for which

$$N(\sin\theta, \cos\theta, V_\pm) = 0 \tag{3.6.45}$$

which in turn involves a squaring operation, so that possibly not all roots of (3.6.44) need satisfy (3.6.45). In fact (only results relevant to situations for which $\alpha > 1$ and $\beta > 1$ are stated here) equation (3.6.45) is satisfied at $\theta = 0$ and π on W_- and at $\theta = \pi/2$ and $3\pi/2$ on W_+. Furthermore, the v_2 displacement is continuous (no jump) at these points. This fact can be observed in figure 53. In a like manner, the w_3 displacement will be continuous on W_- at $\theta = \pi/2$ and $3\pi/2$ and on W_+ at $\theta = 0$ and π.

The wavefront displacement behaviour at an ordinary point on W can now be summarized as follows,

The displacement jump across the wavefront at an ordinary point on $W_\pm$ is singular as $d^{-1/2}$ where d measures distance from the left side of $W_\pm$ as the point in question on $W_\pm$ is approached along the direction of the normal vector. For certain special values of θ, for which $N(\sin\theta, \cos\theta, V_\pm(\theta))$, $\sin\theta\cos\theta$, or $K(\sin\theta, \cos\theta, V_+(\theta))$ vanish, the respective displacements v_2, v_3 or w_3 are continuous across $W_\pm$.

Displacement Singularity at a Cuspidal Point of W

It now remains to investigate the wavefront behaviour in the vicinity of a cuspidal point on W_+ (there are no such points on W_-). Recall from section 2.8 that a cusp point on W_+ corresponds to an inflection point on the outer branch of the slowness curve N_+. At such a point the line L, defined in (3.6.29), will intersect $N_\pm$ in such a way that $f(p|\bar{z}_+, \bar{y}_+)$ will have a triple p root at $p = p_0$. Proceeding as above

$$f(p|z, y) \approx H + f'''(p_0|\bar{z}_+, \bar{y}_+)(p - p_0)^3/3!, \qquad (3.6.46)$$

where

$$f'''(p_0|\bar{z}_+, \bar{y}_+) = \frac{24 \cos\theta \sin^3\theta}{(B^2 - 4A)^2} \, [\gamma - (\alpha\beta + 1)] \, [\gamma - (\alpha + \beta)]$$

$$\times \, [(1 - \beta)^2 \sin^4\theta - (1 - \alpha)^2 \cos^4\theta], \qquad (3.6.47)$$

or alternatively

$$f'''(p_0|\bar{z}_+, \bar{y}_+) = -2 \sin^2\theta \sqrt{B^2 - 4A} \, V_+(V_+' + V_+'''). \qquad (3.6.48)$$

Using the approximation (3.6.46) for f in a displacement function residue calculation will result in an expression proportional to $|H|^{-2/3}$, i.e. in which the displacement is singular as $d^{-2/3}$. Here, as above, d would be measured along the normal direction at the cuspidal point. But now the normal direction is perpendicular to the (limiting) tangent line L (see figure 55) so that 'crossing the wavefront' in this direction actually gives a double crossing with the result that the lacuna has not been penetrated. In order to use an expression like that of equation (3.6.28), the cusp tip must be approached along the line L, since this is the only direction by which the

Figure 55. Lacuna Penetration Direction at a Cuspidal Point.

lacuna can be penetrated. (Note that although $V_+ + V_+'' = 0$ here, $V_+' + V_+''' \neq 0$ so that the limiting slope on W_+ is still $-\cot\theta$.) But for $(\bar{z}, \bar{y})$ on L, $H = 0$ so that a more accurate approximation to $f(p|\bar{z}, \bar{y})$, than that given in (3.6.46) is required. By analyzing the perturbed roots of f, it can be shown that when $(\bar{z}, \bar{y})$ is near a cusp of W_+ and on the line L

$$f(p|\bar{z}, \bar{y}) \approx a_0 + a_1(p - p_0) + \tfrac{1}{2}a_2(p - p_0)^2 + \tfrac{1}{6}a_3(p - p_0)^3, \tag{3.6.49}$$

where

$$a_0 = \tfrac{1}{2}[f_{\bar{z}\bar{z}}(p_0|\bar{z}_+, \bar{y}_+)(\bar{z} - \bar{z}_+)^2 + 2f_{\bar{z}\bar{y}}(p_0|\bar{z}_+, \bar{y}_+)(\bar{z} - \bar{z}_+)(\bar{y} - \bar{y}_+)$$

$$+ f_{\bar{y}\bar{y}}(p_0|\bar{z}_+, \bar{y}_+)(\bar{y} - \bar{y}_+)^2], \tag{3.6.50}$$

$$a_1 = [f_{\bar{z}}'(\bar{z} - \bar{z}_+) + f_{\bar{y}}'(\bar{y} - \bar{y}_+)] = -\frac{2V_+}{\sin^3\theta}\sqrt{B^2 - 4A}(\bar{z} - \bar{z}_+), \tag{3.6.51}$$

$$a_2 = [f_{\bar{z}}''(\bar{z} - \bar{z}_+) + f_{\bar{y}}''(\bar{y} - \bar{y}_+)] \quad \text{and} \quad a_3 = f'''. \tag{3.6.52}$$

In the subsequent analysis a_0, being $O[(\bar{z} - \bar{z}_+)^2]$ can be neglected. It is helpful here to introduce the terms

$$\phi(\theta) = \frac{6a_1}{a_3} = \frac{6(\bar{z} - \bar{z}_+)}{\sin^5\theta(V_+' + V_+'')} \tag{3.6.53}$$

and

$$\lambda(\theta) = \frac{a_2}{2a_1\sin\theta} - \frac{f^{(4)}}{2f^{(3)}\sin\theta}$$

$$= \frac{2[(\alpha + 1)\cos\theta - 2V_+\bar{z}_+]}{\sqrt{B^2 - 4A}} + \frac{\bar{z}_+}{V_+}$$

$$+ \frac{6(\alpha - \bar{z}_+^2)(1 - \bar{z}_+^2)}{\sin^3\theta\sqrt{B^2 - 4A}\,V_+(V_+' + V_+''')}, \tag{3.6.54}$$

it being understood that $\bar{z}$ and $\bar{y}$ lie on L.

The same notation for the displacement that was introduced in (3.6.28) will again be used at a cuspidal point of W_+, with the understanding that $\bar{z}$ and $\bar{y}$ immediately to the left of W_+ should be interpreted as $\bar{z}$ and $\bar{y}$ on L and immediately to the left of (i.e. exterior to) the lacuna. From the observation of typical W_+ and N_+ curves, paying particular attention to the sign of the curvature slope at an inflection point of N_+, the conclusion may be drawn that $\phi(\theta) > 0$ when $\bar{z}$ is on L and immediately to the left of a cusp and $\phi(\theta) < 0$ to the right.

From equation (3.6.49) the perturbed upper half p-plane pole of $f^{-1}(p|\bar{z}, \bar{y})$ is given by

LEFT of W_+ $p = p_0 + i\sqrt{\phi(\theta)}$. \hfill (3.6.55)

RIGHT of W_+ All poles are real

From (3.2.20) a residue calculation now gives

$$v_2|_{\text{LEFT}} = -\frac{1}{\pi\tau} \text{Im} \frac{N(1, p, \bar{y} + \bar{z}p)}{f'(p|\bar{z}, \bar{y})}\bigg|_{p = p_0 + i\sqrt{\phi}}, \tag{3.6.56}$$

$$= \frac{\lambda(\theta) \sin \theta N + N'}{\frac{1}{3}\pi\tau f'''} |\phi|^{-1/2}. \tag{3.6.57}$$

Set

$$g(\theta) = -\frac{1}{3}[2\pi\tau V_+(V_+' + V_+''') \sin^3\theta \sqrt{B^2 - 4A}] \tag{3.6.58}$$

then the jumps for the various displacement components near a cuspidal point of W_+, when $\bar{z}$ is on the line L: $(\bar{z} - \bar{z}_+)\cos\theta + (\bar{y} - \bar{y}_+)\sin\theta = 0$ and just outside the lacuna, are given by

$$[v_2]_+ = \frac{\lambda(\theta)N(\sin\theta, \cos\theta, V_+) + 2(\alpha\cos\theta - \bar{z}_+V_+)}{g(\theta)} |\phi(\theta)|^{-1/2} \tag{3.6.59}$$

$$[v_3]_+ = [w_2]_+ = -\frac{\lambda(\theta)\kappa\sin\theta\cos\theta + \kappa\sin\theta}{g(\theta)} |\phi(\theta)|^{-1/2} \tag{3.6.60}$$

and

$$[w_3]_+ = \frac{\lambda(\theta)K(\sin\theta, \cos\theta, V_+) + 2(\cos\theta - \bar{z}_+V_+)}{g(\theta)} |\phi(\theta)|^{-1/2}. \tag{3.6.61}$$

Equations (3.6.59)–(3.6.61) thus predict that when a cuspidal point of W_+ is approached along L from the exterior side of the lacuna, the displacement jumps are singular like $d^{-1/2}$.

This now completes the somewhat lengthy discussion of the Green's tensor wave-front singularities. The results as stated above are sufficiently general to include all of the 13 crystals listed in table 7. There are, however, a number of mathematically interesting cases such as $\alpha = 1$ (so that N contains double points) and a point lacuna (caused by two inflection points coalescing on N_+) which have been excluded from this section. The point lacuna has been investigated by Payton [41], this paper also having served as the basis for this section.

7. The Stress Field Induced by a Line Load

In this section the stress field caused by a uniform line load acting along the x-axis, will be derived. Because of the uniformity of loading in the x-direction, the u component of displacement, together with derivatives of the other displacement components in this direction, may be assumed to vanish. From (1.3.2) the stress-strain relations reduce to

$$\sigma_{11} = c_{12}\frac{\partial v}{\partial y} + c_{13}\frac{\partial w}{\partial z}, \qquad \sigma_{22} = c_{11}\frac{\partial v}{\partial y} + c_{13}\frac{\partial w}{\partial z},$$

$$\sigma_{33} = c_{13}\frac{\partial v}{\partial y} + c_{33}\frac{\partial w}{\partial z}, \qquad \sigma_{23} = c_{44}\left(\frac{\partial v}{\partial z} + \frac{\partial w}{\partial y}\right), \tag{3.7.1}$$

$$\sigma_{13} = 0 \quad \text{and} \quad \sigma_{12} = 0.$$

By using the dynamic equilibrium equations, the stress components may be uncoupled to give

$$Q[\sigma_{11}^{(2)}] = \frac{\partial}{\partial y}M_{112}[g] \quad \text{where} \quad M_{112} = \left[-(\beta - 2\delta)N + \kappa(\kappa - 1)\frac{\partial^2}{\partial z^2}\right],$$

$$\tag{3.7.2}$$

$$Q[\sigma_{11}^{(3)}] = \frac{\partial}{\partial z}M_{113}[g] \quad \text{where} \quad M_{113} = \left[(\beta - 2\delta)\kappa\frac{\partial^2}{\partial y^2} - (\kappa - 1)K\right],$$

$$\tag{3.7.3}$$

$$Q[\sigma_{2}^{(2)}] = \frac{\partial}{\partial y}M_{222}[g] \quad \text{where} \quad M_{222} = \left[-\beta N + \kappa(\kappa - 1)\frac{\partial^2}{\partial z^2}\right], \tag{3.7.4}$$

$$Q[\sigma_{22}^{(3)}] = \frac{\partial}{\partial z}M_{223}[g] \quad \text{where} \quad M_{223} = \left[\beta\kappa\frac{\partial^2}{\partial y^2} - (\kappa - 1)K\right], \tag{3.7.5}$$

$$Q[\sigma_{33}^{(2)}] = \frac{\partial}{\partial y}M_{332}[g] \quad \text{where} \quad M_{332} = \left[-(\kappa - 1)N + \alpha\kappa\frac{\partial^2}{\partial z^2}\right], \tag{3.7.6}$$

$$Q[\sigma_{33}^{(3)}] = \frac{\partial}{\partial z}M_{333}[g] \quad \text{where} \quad M_{333} = \left[\kappa(\kappa - 1)\frac{\partial^2}{\partial y^2} - \alpha K\right], \tag{3.7.7}$$

$$Q[\sigma_{23}^{(2)}] = \frac{\partial}{\partial z}M_{232}[g] \quad \text{where} \quad M_{232} = \left[-N + \kappa\frac{\partial^2}{\partial y^2}\right], \tag{3.7.8}$$

$$Q[\sigma_{23}^{(3)}] = \frac{\partial}{\partial y}M_{233}[g] \quad \text{where} \quad M_{233} = \left[\kappa\frac{\partial^2}{\partial z^2} - K\right]. \tag{3.7.9}$$

Here the material constant δ has been defined in equation (1.5.22) while

$$g(y, z, t) = \delta(y)\delta(z)H(\tau). \tag{3.7.10}$$

Note that the line load time dependence for the stress problem is $H(t)$, whereas for the displacement problem of section 3.1 it was $\delta(t)$. This change is to facilitate ease in using the integral transform inversion technique. A time derivative of the stress solutions obtained below would correspond to a $\delta(t)$ load time dependence. Also the step function loading will result (as $t \to +\infty$) in a static stress distribution as a special case of the solution.

The stress components in equation (3.7.1) and the superscripted stress components of equations (3.7.2)–(3.7.10) are related by

$$\sigma_{11} = \rho a_2 \sigma_{11}^{(2)} + \rho a_3 \sigma_{11}^{(3)}, \qquad \sigma_{22} = \rho a_2 \sigma_{22}^{(2)} + \rho a_3 \sigma_{22}^{(3)},$$

$$\sigma_{33} = \rho a_2 \sigma_{33}^{(2)} + \rho a_3 \sigma_{33}^{(3)} \quad \text{and} \quad \sigma_{23} = \rho a_2 \sigma_{23}^{(2)} + \rho a_3 \sigma_{23}^{(3)}, \tag{3.7.11}$$

As in section 3.2, expressions for the various stress components can be reduced to a residue calculation. Because of the similarities, only two stress components will be listed here.

$$\pi \tau \sigma_{11}^{(2)} = -\mathrm{Im} \sum_p \left\{ \text{residues of } \frac{M_{112}(1, p, \bar{y} + \bar{z}p)}{(\bar{y} + \bar{z}p)Q(1, p, \bar{y} + \bar{z}p)} \right.$$

$$\left. \text{in the upper half } p\text{-plane} \right\}, \tag{3.7.12}$$

and

$$\pi \tau \sigma_{11}^{(3)} = -\mathrm{Im} \sum_p \left\{ \text{residues of } \frac{p M_{113}(1, p, \bar{y} + \bar{z}p)}{(\bar{y} + \bar{z}p)Q(1, p, \bar{y} + \bar{z}p)} \right.$$

$$\left. \text{in the upper half } p\text{-plane} \right\}. \tag{3.7.13}$$

The stresses along the z-axis may conveniently be expressed in terms of the displacement functions $V(\bar{z})$ and $W(\bar{z})$ of equation (3.5.8). The four nonvanishing stress components are

$$\pi z \sigma_{11}^{(3)}(0, z, \tau) = -\kappa (\beta - 2\delta) S(\bar{z}) - (\kappa - 1) W(\bar{z}), \tag{3.7.14}$$

$$\pi z \sigma_{22}^{(3)}(0, z, \tau) = \beta \kappa S(\bar{z}) - (\kappa - 1) W(\bar{z}), \tag{3.7.15}$$

$$\pi z \sigma_{33}^{(3)}(0, z, \tau) = \kappa (\kappa - 1) S(\bar{z}) - \alpha W(\bar{z}), \tag{3.7.16}$$

and

$$\pi z \sigma_{23}^{(3)}(0, z, \tau) = \kappa S(\bar{z}) - V(\bar{z}), \tag{3.7.17}$$

where

$$S(\bar{z}) = \frac{(1 - \bar{z}^2) V(\bar{z}) - (\alpha - \bar{z}^2) W(\bar{z})}{(1 - \alpha\beta) + (\beta - 1)\bar{z}^2}. \tag{3.7.18}$$

Note that the static solution (along the z-axis) is given by e.g.,

$$\pi z \sigma_{11}^{(3)}(0, z, \infty) = -\kappa (\beta - 2\delta) S(0) - (\kappa - 1) W(0). \tag{3.7.19}$$

By taking the limit of e.g., equation (3.7.12) as $\tau \to +\infty$, before the residue calculation has been performed, it would not be difficult to obtain closed form expressions for the various static stress components at any arbitrary field point (y, z).

The analysis of this section follows closely that given by the author in [34].

8. Remarks on the Green's Tensor Representation When the Line Load is Not Perpendicular to the Material Symmetry Axis

The line load problem of section 3.1 will now be generalized so that the loading axis is not normal to the symmetry axis (z-axis). Let $Ox_1x_2x_3$ replace the previous co-ordinate frame $Oxyz$ so that the symmetry axis is now the x_3-axis. Further, let $Ox_1'x_2'x_3'$ denote a second coordinate frame such that the x_1'-axis coincides with the line load. With respect to the primed coordinate system, the applied load (per unit mass) is

$$f_i' = (0, a_2', a_3')\delta(x_2')\delta(x_3')\delta(t). \tag{3.8.1}$$

The equations of motion (1.1.6), in the unprimed frame are

$$c_{ijkl}\frac{\partial^2 u_k}{\partial x_l \partial x_j} + \rho f_i = \rho \frac{\partial^2 u_i}{\partial t^2} \tag{3.8.2}$$

where

$$f_i = (\alpha_{21}a_2' + \alpha_{31}a_3', \alpha_{22}a_2' + \alpha_{32}a_3', \alpha_{23}a_2' + \alpha_{33}a_3')\delta(x_2')\delta(x_3')\delta(t). \tag{3.8.3}$$

Here the α_{ij}'s have been defined in equation (1.2.1). The loading for the equations of motion (3.8.2), in an obvious notation, thus has the form

$$f_i = (a_1, a_2, a_3)f(x_1, x_2, x_3, t), \tag{3.8.4}$$

where

$$f(x_1, x_2, x_3, t) = \delta(x_2')\delta(x_3')\delta(t), \tag{3.8.5}$$

with

$$x_2' = \alpha_{21}x_1 + \alpha_{22}x_2 + \alpha_{23}x_3 \quad \text{and} \quad x_3' = \alpha_{31}x_1 + \alpha_{32}x_2 + \alpha_{33}x_3 \tag{3.8.6}$$

The above problem can now be placed in the form (1.7.3) for a three-dimensional problem. However, now the partial differential equations will be (for certain components of the Green's tensor) of sixth order. The left hand side operator will be the product of Q and a second order operator L. But because of the special way in which the three coordinates x_1, x_2 and x_3 enter into the loading (3.8.5) through the two combinations (3.8.6), a solution (e.g., for u_1) may be sought in the form

$$u_1(x_1, x_2, x_3, t) = \bar{u}_1(x_2', x_3', t). \tag{3.8.7}$$

Then using $\partial u_1/\partial x_i = (\partial \bar{u}_1/\partial x_k')(\partial x_k'/\partial x_i) = \alpha_{ki}\partial \bar{u}_1/\partial x_k'$ etc., for higher derivatives, the problem again reduces to one of plane strain in the variables x_2', x_3' and t. Using the primed variables causes the L and Q operators to become more extended in that they will now contain a richer variety of mixed derivatives, but Q will remain a fourth order, linear, homogeneous, partial differential operator with constant coefficients. The transform inversion technique of section 3.2 is still applicable and hence the arbitrarily oriented line load problem can thus be solved explicitly.

PART II. THREE SPACE DIMENSIONS

9. Integral Transform Representation of the Displacement Field

The central problem to be studied in the remaining sections of this chapter has been formulated in section 1.7. The excitation of the undisturbed solid is caused by the sudden application of a point body force at the origin with arbitrary orientation. For reasons of mathematical convenience (associated with the three-dimensional nature of the problem) the time dependence of the loading is taken to be a unit step $H(t)$ rather than $\delta(t)$. A time derivative of the displacement components given below would give the corresponding displacements for a $\delta(t)$ time loading dependence and thus establish the three-dimensional Green's tensor. Another reason for preferring the $H(t)$ time loading is that the static solution (and hence static Green's tensor) may be expected after a sufficient (and finite) time interval has elapsed.

An examination of equations (1.7.3)–(1.7.5) reveals certain symmetries among the nine displacement components so that it is only necessary to calculate the four quantities u_1, u_2, u_3 and w_3 since

$$v_1(x, y, z, \tau) = u_2(x, y, z, \tau), \qquad v_2(x, y, z, \tau) = u_1(y, x, z, \tau)$$

$$w_1(x, y, z, \tau) = u_3(x, y, z, \tau), \qquad v_3(x, y, z, \tau) = u_3(y, x, z, \tau)$$

and

$$w_2(x, y, z, \tau) = v_3(x, y, z, \tau). \tag{3.9.1}$$

Integral solutions to equations (1.7.3) can now be obtained by a straightforward application of Fourier and Laplace transforms. Denote the triple Fourier exponential transform (in x, y and z) of u_1 by u_1^* and the Laplace transform (in τ) by $\bar{u}_1$. Then

$$\bar{u}_1^*(\xi, \eta, \zeta, s) = \int_0^\infty \int_{-\infty}^{+\infty} \int_{-\infty}^{+\infty} \int_{-\infty}^{+\infty} u_1(x, y, z, \tau) e^{-i(\xi x + \eta y + \zeta z) - s\tau} \, dx \, dy \, dz \, d\tau \tag{3.9.2}$$

with inverse

$$u_1(x, y, z, \tau) = \frac{1}{(2\pi)^4 i} \int_c \int_{-\infty}^{+\infty} \int_{-\infty}^{+\infty} \int_{-\infty}^{+\infty} \bar{u}_1^*(\xi, \eta, \zeta, s) e^{i(\xi x + \eta y + \zeta z) + s\tau} \, d\xi \, d\eta \, d\zeta \, ds.$$

$$\tag{3.9.3}$$

Of course similar expressions hold true for the other relevant displacement components u_2, u_3, and w_3. As typical of the group of displacement transforms (and because they will be studied in detail below) $\bar{u}_1^*$ and $\bar{w}_3^*$ will be listed here,

$$\bar{u}_1^* = \frac{-1}{sL(i\xi, i\eta, i\zeta, s)} + \frac{\xi^2 M(i\xi, i\eta, i\zeta, s)}{sL(i\xi, i\eta, i\zeta, s)Q(i\xi, i\eta, i\zeta, s)}, \tag{3.9.4}$$

and

$$\bar{w}_3^* = \frac{-K(i\xi, i\eta, i\zeta, s)}{sQ(i\xi, i\eta, i\zeta, s)}. \tag{3.9.5}$$

95

10. Transform Inversion for the Special Case $\gamma = \alpha + \beta$

For certain special combinations of the elastic parameters α, β and γ the fourth order partial differential operator Q will factor into two second order operators. According to the classification of Chapter 2, this will occur under cases 2, 3, 5, 6, 7, 8 and 10 for $\gamma = 1 + \alpha\beta$ and under cases 1, 2, 3, 4, 7, 8 and 10 for $\gamma = \alpha + \beta$. Either of these constraints causes an enormous simplification in the mathematics of the Green's tensor construction. With the constraint, it is possible to give a closed form solution to the problem, whereas without the constraint (at least up to the present date) explicit results have only been obtained along the symmetry axis. In the present section the constraint $\gamma = \alpha + \beta$ will be imposed on the solid with the further assumption that $\beta > \alpha > 1$ so as to stay within the bounds of case 4, since this case contains the hexagonal crystals listed in Table 7. The assumption $\beta > \alpha > 1$ is not necessary to the analysis, in fact by analytic continuation (see Payton [42]), the results of this section may be easily extended to the situation $\alpha > \beta > 1$.

The mathematical simplification resulting from the constraint $\gamma = \alpha + \beta$ has been appreciated by Carrier [43] and Cameron and Eason [44]. From a physical stand-point, there exists little experimental evidence to support the reality of such a solid. However, the value of such a constraint relation lies not in its approximation to any particular material, but rather in the simplification of the problem while still pre-serving some of the qualitative features of wave propagation in transversely isotropic media. Such a solution can also serve as a check case for asymptotic results obtained from more complicated problems. Incidentally, the isotropic Green's tensor may be obtained from the results presented in this section by setting $\delta = 1$ and taking a limit as $\beta \to \alpha + 0$.

The displacement field produced in an isotropic solid of infinite extent, by a time-dependent point load is generally attributed to Stokes [45], although certain features of the solution were anticipated earlier by Poisson [46]. Stokes noticed that the dilation (divergence of the displacement vector) and rotation (curl of the displacement vector), in the absence of body forces, separately satisfy wave equations with different propagation speeds. After obtaining integral solutions to these equations for the dilation and rotation, Stokes indicated how the displacement components could then be found.

Some of the details will now be given for the inversion of the transformed displacement $\bar{u}_1^*$ of equation (3.9.4). The resourceful reader will be able to duplicate the steps for the other components of the Green's tensor. Write equation (3.9.4) as

$$\bar{u}_1^* = \bar{I}_1^* + \bar{I}_2^*, \tag{3.10.1}$$

with

$$\bar{I}_1^* = -\frac{1}{sL} \quad \text{and} \quad \bar{I}_2^* = \frac{\xi^2 M}{sLQ} \tag{3.10.2}$$

where, from (1.5.20), (1.5.21), (1.7.5) and using $\gamma = \alpha + \beta$

$$L(i\xi, i\eta, i\zeta, s) = -[\zeta^2 + \delta(\xi^2 + \eta^2) + s^2],$$

$$M(i\xi, i\eta, i\zeta, s) = -[(1 - \alpha - \beta + \alpha\delta)\zeta^2 + (\delta - \beta)(\xi^2 + \eta^2) + (\delta - \beta)s^2]$$

$$(3.10.3)$$

and

$$Q(i\xi, i\eta, i\zeta, s) = [\alpha\zeta^2 + \beta(\xi^2 + \eta^2) + s^2][\zeta^2 + (\xi^2 + \eta^2) + s^2].$$

By suitably stretching the integration variables ξ and η, and space variables x and y, the operator L becomes the isotropic wave operator (i.e. with δ replaced by 1), whose inverse is well known. Thus

$$I_1 = \frac{H(\tau - R_0)}{4\pi\delta R_0} \tag{3.10.4}$$

where

$$R_0 = (r^2/\delta + z^2)^{1/2}, \qquad r = (x^2 + y^2)^{1/2} \tag{3.10.5}$$

and H denotes the Heaviside unit step function.

The complete inversion of $\bar{I}_2^*$ is much harder. First perform the ζ-inversion

$$\frac{1}{2\pi} \int_{-\infty}^{+\infty} \bar{I}_2^* \, e^{i\zeta z} \, d\zeta. \tag{3.10.6}$$

This integration is easy since the integrand contains only simple pole singularities in the complex ζ-plane. After the ζ-inversion, and with the help of (3.9.3) the Laplace transform of I_2 may be written

$$\bar{I}_2 = \frac{1}{2(2\pi)^2} \int_{-\infty}^{+\infty}\int_{-\infty}^{+\infty} e^{i(\xi x + \eta y)} d\xi d\eta \left\{ -\frac{\xi^2 \exp\left[-|z|\{\delta(\xi^2 + \eta^2) + s^2\}^{1/2}\right]}{\{\delta(\xi^2 + \eta^2) + s^2\}^{1/2}(\xi^2 + \eta^2)} \right.$$

$$-\frac{(\beta - 1)\xi^2 \exp\left[-|z|\{\beta(\xi^2 + \eta^2)/\alpha + s^2/\alpha\}^{1/2}\right]}{\{\beta(\xi^2 + \eta^2)/\alpha + s^2/\alpha\}^{1/2}[(\alpha - \beta)(\xi^2 + \eta^2) + (\alpha - 1)s^2]}$$

$$\left. +\frac{(\alpha - 1)\xi^2\{(\xi^2 + \eta^2) + s^2\}^{1/2} \exp\left[-|z|\{(\xi^2 + \eta^2) + s^2\}^{1/2}\right]}{(\xi^2 + \eta^2)[(\alpha - \beta)(\xi^2 + \eta^2) + (\alpha - 1)s^2]} \right\}. \tag{3.10.7}$$

Next rearrange the integrand, changing to polar coordinates (r, ψ) for the space variables x and y, and (ρ, θ) for the integration variables ξ and η by

$$x = r \cos \psi, \qquad y = r \sin \psi, \tag{3.10.8}$$

and

$$\xi = \rho s \cos (\theta + \psi), \qquad \eta = \rho s \sin (\theta + \psi). \tag{3.10.9}$$

Then, after some intermediate calculations, and the θ-integration variable change

$$\cos \theta = p, \tag{3.10.10}$$

it follows that

$$\bar{I}_2 = \frac{1}{2\pi^2} \int_0^1 (p^2 \cos^2 \psi + (1 - p^2) \sin^2 \psi) \frac{dp}{\sqrt{1 - p^2}} [-\bar{H}_1 - \bar{H}_2 + \bar{H}_3], \tag{3.10.11}$$

where

$$\bar{H}_1 = \operatorname{Re}\int_0^\infty \frac{\rho\, d\rho}{\sqrt{\delta\rho^2 + 1}}\, \exp\{s[irp\rho - |z|\sqrt{\delta\rho^2 + 1}]\}, \tag{3.10.12}$$

$$\bar{H}_2 = \operatorname{Re}\int_0^\infty \frac{(\beta - 1)\rho^3\, d\rho}{\sqrt{\frac{\beta}{\alpha}\rho^2 + \frac{1}{\alpha}}\,[(\alpha - \beta)\rho^2 + (\alpha - 1)]}\, \exp\left\{s\left[irp\rho - |z|\sqrt{\frac{\beta}{\alpha}\rho^2 + \frac{1}{\alpha}}\right]\right\} \tag{3.10.13}$$

and

$$\bar{H}_3 = \operatorname{Re}\int_0^\infty \frac{(\alpha - 1)\sqrt{\rho^2 + 1}\,\rho\, d\rho}{[(\alpha - \beta)\rho^2 + (\alpha - 1)]}\, \exp\{s[irp\rho - |z|\sqrt{\rho^2 + 1}]\}. \tag{3.10.14}$$

Note that separately, the integrals $\bar{H}_2$ and $\bar{H}_3$ have a simple pole singularity on the ρ-plane integration path since $(\alpha - 1) > 0$ while $(\alpha - \beta) < 0$. However collectively, the combination $-\bar{H}_2 + \bar{H}_3$ gives a zero residue contribution at $\rho = [(\alpha - 1)/(\beta - \alpha)]^{1/2}$. For this reason, integrals $\bar{H}_2$ and $\bar{H}_3$ may be interpreted as Cauchy principal-value integrals.

The method of Cagniard [47] and Pekeris [48] will now be invoked to deform the ρ-integration path in such a way that the exponential terms encountered in (3.10.12)–(3.10.14) are real. In order to illustrate the method, without introducing excessive details, only that part of I_2 associated with H_1 will be explicitly derived. Toward this goal, set

$$\rho = \delta^{-1/2}\eta \quad \text{and} \quad rp\delta^{-1/2} = a \tag{3.10.15}$$

so that (3.10.12) may be written

$$\delta\bar{H}_1 = \operatorname{Re}\int_0^\infty \frac{e^{sg(\eta)}\eta\, d\eta}{\sqrt{\eta^2 + 1}}, \tag{3.10.16}$$

where

$$g(\eta) = ia\eta - |z|\sqrt{\eta^2 + 1}. \tag{3.10.17}$$

From (3.10.15), note that $a \geqslant 0$ is a constant, insofar as the η-integration is concerned.

Look now for a path Γ in the complex η-plane ($\eta = \eta_1 + i\eta_2$) where

$$g(\eta) = -T = \text{real}, \tag{3.10.18}$$

hence

$$\operatorname{Im} g(\eta) = 0 \quad \text{on} \quad \Gamma. \tag{3.10.19}$$

Define the function $\sqrt{\eta^2 + 1}$ in the cut η-plane as

$$\sqrt{\eta^2 + 1} = (R_1 R_2)^{1/2}\, e^{i(\theta_1 + \theta_2)/2} \tag{3.10.20}$$

where

$$R_1 = |\eta - i|, \qquad R_2 = |\eta + i|, \qquad -\pi < \theta_1 \leqslant \pi, \qquad -\pi < \theta_2 \leqslant \pi. \tag{3.10.21}$$

Figure 56 shows the upper half η-plane portion of this cut. By solving (3.10.18) for η, it is easily deduced that

$$\eta = \frac{|z|}{R} \sqrt{T^2 - R^2} + i\frac{aT}{R^2} \quad \text{with} \quad R \leqslant T < \infty \qquad (3.10.22)$$

on the first quadrant portion of Γ. Here

$$R = \sqrt{z^2 + a^2}. \qquad (3.10.23)$$

Recalling that $\eta = \eta_1 + i\eta_2$ and eliminating T from (3.10.22) shows that Γ is an arc of the hyperbola

$$-\frac{\eta_1^2}{z^2/R^2} + \frac{\eta_2^2}{a^2/R^2} = 1, \qquad (3.10.24)$$

with the asymptote

$$|z|\eta_2 = a\eta_1. \qquad (3.10.25)$$

Now on Γ

$$\frac{d\eta}{\sqrt{\eta^2 + 1}} = \frac{dT}{\sqrt{T^2 - R^2}}. \qquad (3.10.26)$$

so that

$$\text{Re} \int_\Gamma \frac{e^{sg(\eta)}\eta \, d\eta}{\sqrt{\eta^2 + 1}} = \text{Re} \int_\infty^R \frac{e^{-sT} dT}{\sqrt{T^2 - R^2}} \left\{ i\frac{aT}{R^2} + \frac{|z|}{R^2} \sqrt{T^2 - R^2} \right\}$$

$$= -\frac{|z|}{R^2} \int_R^\infty e^{-sT} dT = -\frac{|z|}{sR^2} e^{-sR}. \qquad (3.10.27)$$

If the integrand of (3.10.16) is now integrated about the closed contour indicated in figure 56, where C is a circular arc, then

$$\int_0^\infty \frac{e^{sg(\eta)}\eta \, d\eta}{\sqrt{\eta^2 + 1}} + \int_C + \int_\Gamma + \int_{a/R}^0 \frac{[\exp\{sg(\eta_2 e^{i\pi/2})\}] i\eta_2 \, d\eta_2}{\sqrt{1 - \eta_2^2}} = 0 \qquad (3.10.28)$$

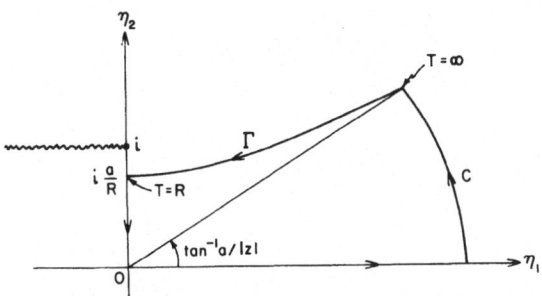

Figure 56. Complex η-Plane Showing the Contours used in Evaluating H_1.

since the interior of the closed contour is free of singularities. Now

$$\int_c \to 0 \quad \text{as} \quad |\eta| \to \infty$$

so that, in view of equations (3.10.16), (3.10.27) and (3.10.28)

$$\delta \bar{H}_1 = \frac{|z|}{sR^2} e^{-sR} - \int_0^{a/R} \frac{[\exp\{-s(a\eta_2 + |z|\sqrt{1-\eta_2^2})\}]\,\eta_2\,d\eta_2}{\sqrt{1-\eta_2^2}} \tag{3.10.29}$$

At this stage it is not apparent that any progress has been made, since expression (3.10.29) for $\bar{H}_1$ is scarcely any simpler than (3.10.12). However the integrand in (3.10.29) is now real and this justifies the labor expended. Next set

$$h = a\eta_2 + |z|\sqrt{1-\eta_2^2}, \tag{3.10.30}$$

then, since $h(\eta_2)$ is monotonic increasing for $0 \leqslant \eta_2 \leqslant a/R$, equation (3.10.30) may be inverted for η_2 giving

$$\eta_2 = \frac{a}{R^2} h - \frac{|z|}{R^2}\sqrt{R^2 - h^2} \tag{3.10.31}$$

with $|z| < h < R$ corresponding to the integration range in (3.10.29). From (3.10.30) and (3.10.31)

$$\frac{\eta_2\,d\eta_2}{\sqrt{1-\eta_2^2}} = \frac{1}{R^2}\left(-|z| + \frac{ah}{\sqrt{R^2-h^2}}\right) dh, \tag{3.10.32}$$

so that the integral in (3.10.29) becomes

$$\int_0^{a/R} \frac{[\exp\{-s(a\eta_2 + |z|\sqrt{1-\eta_2^2})\}]\,\eta_2\,d\eta_2}{\sqrt{1-\eta_2^2}}$$

$$= \int_{|z|}^R e^{-sh}\left[\left(-|z| + \frac{ah}{\sqrt{R^2-h}}\right)\Big/ R^2\right] dh$$

$$= -\frac{|z|}{R^2}\left[-\frac{e^{-sh}}{s}\right]_{|z|}^R + \frac{a}{R^2}\int_{|z|}^R \frac{e^{-sh}h\,dh}{\sqrt{R^2-h^2}}. \tag{3.10.33}$$

Replacing the integral in (3.10.29) by the expression (3.10.33) gives

$$\delta \bar{H}_1 = \frac{|z|}{sR^2} e^{-s|z|} - \frac{a}{R^2}\int_{|z|}^R \frac{e^{-sh}h\,dh}{\sqrt{R^2-h^2}}. \tag{3.10.34}$$

The Laplace inversion of (3.10.34) is now immediate (this being the point of the method of Cagniard and Pekeris)

$$\delta H_1 = \frac{|z|}{R^2} H(\tau - |z|) - \frac{a}{R^2}\int_{|z|}^R \frac{\delta(\tau - h)h\,dh}{\sqrt{R^2-h^2}} \tag{3.10.35}$$

or

$$\delta H_1 = \frac{|z|}{R^2} H(\tau - |z|) - \frac{a\tau/R^2}{\sqrt{R^2-\tau^2}} H(\tau - |z|)H(R - \tau). \tag{3.10.36}$$

The integrals $\bar{H}_2$ and $\bar{H}_3$ given in (3.10.13) and (3.10.14) can be (with additional effort) inverted in much the same way as $\bar{H}_1$. Thus at this stage, from (3.10.11)

$$I_2 = \frac{1}{2\pi^2} \int_0^1 (p^2 \cos^2 \psi + (1 - p^2) \sin^2 \psi) \frac{dp}{\sqrt{1 - p^2}} (-H_1 - H_2 + H_3).$$

(3.10.37)

Call

$$I_{21} = \int_0^1 (p^2 \cos^2 \psi + (1 - p^2) \sin^2 \psi) \frac{dp}{\sqrt{1 - p^2}} (-H_1).$$ (3.10.38)

The evaluation of this integral will serve to illustrate how the remaining two integrals in (3.10.37) involving H_2 and H_3 could be handled.

From (3.10.38), using the formula for R from (3.10.23),

$$I_{21} = -\frac{|z|}{\delta} H(\tau - |z|) \int_0^1 \frac{p^2 \cos^2 \psi + (1 - p^2) \sin^2 \psi}{\frac{r^2}{\delta} p^2 + z^2} \frac{dp}{\sqrt{1 - p^2}}$$

$$+ \frac{\tau}{\delta} H(\tau - |z|) \int_0^1 \frac{p^2 \cos^2 \psi + (1 - p^2) \sin^2 \psi}{R^2 \sqrt{R^2 - \tau^2}} \frac{r\delta^{-1/2} p \, dp}{\sqrt{1 - p^2}} H(R - \tau),$$

(3.10.39)

or

$$I_{21} = -\frac{|z|}{\delta} H(\tau - |z|)A + \frac{\tau}{\delta} H(\tau - |z|)B$$ (3.10.40)

in an obvious notation for the integrals A and B. In the integrand of A set $p = \sin \theta$ so that

$$A = \int_0^{\pi/2} \frac{\sin^2 \theta \cos^2 \psi + \cos^2 \theta \sin^2 \psi}{r^2 \delta^{-1} \sin^2 \theta + z^2} d\theta$$

$$= \int_0^{\pi/2} \frac{\sin^2 \theta \cos^2 \psi + \cos^2 \theta \sin^2 \psi}{R_0^2 \sin^2 \theta + z^2 \cos^2 \theta} d\theta,$$ (3.10.41)

where

$$R_0 = (r^2 \delta^{-1} + z^2)^{1/2}.$$ (3.10.42)

Dividing the numerator and the denominator of the integrand in (3.10.41) by $\sin^2 \theta$ gives

$$A = \int_0^{\pi/2} \frac{\cos^2 \psi + \sin^2 \psi \cot^2 \theta}{R_0^2 + z^2 \cot^2 \theta} d\theta,$$ (3.10.43)

After making the substitution $\cot \theta = q$ and noting that the resultant integrand is an even function of q,

$$A = \frac{1}{2} \int_{-\infty}^{+\infty} \frac{(\cos^2 \psi + q^2 \sin^2 \psi) dq}{(R_0^2 + z^2 q^2)(1 + q^2)}.$$ (3.10.44)

This latter integral is easily evaluated by residues

$$A = \pi i \left[\frac{\cos^2 \psi - \dfrac{R_0^2}{z^2} \sin^2 \psi}{2iz^2 \dfrac{R_0}{|z|} \left(1 - \dfrac{R_0^2}{z^2}\right)} + \frac{\cos^2 \psi - \sin^2 \psi}{(R_0^2 - z^2)2i} \right],$$

(3.10.45)

or

$$A = \frac{\pi \delta}{2|z|} \left[\frac{(R_0 - |z|) \sin^2 \psi}{r^2} + \frac{|z|(R_0 - |z|) \cos^2 \psi}{R_0 r^2} \right].$$

(3.10.46)

In the integral B, change the integration variable from p to R, thus

$$B = \int_{|z|}^{R_0} \frac{\dfrac{\delta}{r^2}(R^2 - z^2)\cos^2 \psi + \left\{1 - \dfrac{\delta}{r^2}(R^2 - z^2)\right\} \sin^2 \psi}{R^2 \sqrt{R^2 - r^2}} \frac{R\,dR}{\sqrt{R_0^2 - R^2}} H(R - r).$$

(3.10.47)

If $\tau < |z|$, the $H(\tau - |z|)$ term in front of B is zero. On the other hand if $\tau > R_0$ then (since R does not exceed R_0 in the integrand of (3.10.47)) it follows that $(R - \tau) < 0$ so that $H(R - \tau) = 0$. Effectively then, the integral B can be written as

$$B = H(R_0 - \tau) \int_{\tau}^{R_0} \frac{\dfrac{\delta}{r^2}(R^2 - z^2)\cos^2 \psi + \left\{1 - \dfrac{\delta}{r^2}(R^2 - z^2)\right\} \sin^2 \psi}{R^2}$$

$$\times \frac{R\,dR}{\sqrt{R^2 - r^2}\,\sqrt{R_0^2 - R^2}}.$$

(3.10.48)

The substitution $(R_0^2 - R^2) = (R_0^2 - r^2)\cos^2 \theta$ converts

$$\frac{R\,dR}{\sqrt{R^2 - r^2}\,\sqrt{R_0^2 - R^2}} \quad \text{to} \quad d\theta$$

(3.10.49)

so that the B integral becomes

$$B = H(R_0 - \tau) \int_0^{\pi/2} \left(\frac{\delta}{r^2}\cos^2 \psi - \frac{\delta}{r^2}\sin^2 \psi \right) d\theta$$

$$+ H(R_0 - \tau) \frac{\delta}{r^2}(-z^2 \cos^2 \psi + R_0^2 \sin^2 \psi) \int_0^{\pi/2} \frac{d\theta}{R_0^2 \sin^2 \theta + \tau^2 \cos^2 \theta}$$

(3.10.50)

The integrand in the first integral is constant, while the second integral is easily shown to be $\pi/(2R_0\tau)$ so

$$B = H(R_0 - \tau) \frac{\pi \delta}{2r^2} \left[\left(1 - \frac{z^2}{R_0\tau}\right)\cos^2 \psi + \left(-1 + \frac{R_0}{\tau}\right)\sin^2 \psi \right].$$

(3.10.51)

Thus from (3.10.40), (3.10.46) and (3.10.51), the expression for I_{21} is obtained,

$$I_{21} = -\frac{\pi}{2} H(\tau - |z|) \left[\frac{R_0 - |z|}{r^2} \sin^2 \psi + \frac{|z|(R_0 - |z|)}{R_0 r^2} \cos^2 \psi \right]$$

$$+ \frac{\pi}{2} H(\tau - |z|) H(R_0 - \tau) \left[\frac{R_0 \tau - z^2}{r^2 R_0} \cos^2 \psi + \frac{R_0 - \tau}{r^2} \sin^2 \psi \right] \quad (3.10.52)$$

Using $H(R_0 - \tau) = 1 - H(\tau - R_0)$ and $H(\tau - |z|) H(\tau - R_0) = H(\tau - R_0)$ allows I_{21} to be written as

$$I_{21} = \frac{\pi(\tau - |z|)(\cos^2 \psi - \sin^2 \psi)}{2r^2} H(\tau - |z|)$$

$$- \frac{\pi [(R_0 \tau - z^2) \cos^2 \psi + R_0(R_0 - \tau) \sin^2 \psi]}{2r^2 R_0} H(\tau - R_0) \quad (3.10.53)$$

The term depending on the step function $H(\tau - |z|)$, if included in the final expression for the u_1 displacement, would violate the known propagational features of a hyperbolic partial differential equation with a point source. In fact this term will disappear when combined with the contributions to u_1 from H_2 and H_3.

Some of the flavor used to obtain the Green's tensor, when $\gamma = \alpha + \beta$ has been indicated by the above analysis of this section. It remains now but to record the expressions for the four displacement components u_1, u_2, u_3 and w_3, which in view of equations (3.9.1) will furnish the complete Green's tensor. To simplify matters, the following terms are used

$$m = \tau - |z|(1 + b^2)^{1/2}, \qquad n = \tau + |z|(1 + b^2)^{1/2},$$
$$M = (r^2 b^2 + m^2)^{1/2}, \qquad N = (r^2 b^2 + n^2)^{1/2},$$
$$R_0 = (r^2/\delta + z^2)^{1/2}, \qquad R_1 = (r^2/\beta + z^2/\alpha)^{1/2},$$
$$R_2 = (r^2 + z^2)^{1/2}, \qquad r = (x^2 + y^2)^{1/2},$$
$$b^2 = (\alpha - 1)/(\beta - \alpha). \quad (3.10.54)$$

Because of the parameter restriction $\beta > \alpha > 1$, the quantity b^2 is positive. The various displacements are now given by

$$u_1(x, y, z, \tau) = \frac{1}{4\pi r^4} \left[\frac{r^4}{\delta R_0} H(\tau - R_0) - \frac{(R_0 \tau - z^2)x^2 + y^2 R_0(R_0 - \tau)}{R_0} \right.$$

$$\times H(\tau - R_0) + \alpha^{1/2}(1 + b^2) \frac{y^2 R_1^2 - x^2 z^2/\alpha}{R_1} H(\tau - R_1)$$

$$+ \left\{ \tau(x^2 - y^2) - \frac{b^2 y^2 R_2^2 - b^2 x^2 z^2}{R_2} \right\} H(\tau - R_2)$$

$$+ \frac{(1 + b^2)^{1/2}}{2} \left\{ \frac{y^2 r^2 b^2 - m^2(x^2 - y^2)}{M} \right.$$

$$+ \frac{y^2 r^2 b^2 - n^2(x^2 - y^2)}{N} \Bigg\} \times \{H(\tau - R_2) - H(\tau - R_1)\} \Bigg],$$

(3.10.55)

$$u_2(x,y,z,\tau) = \frac{xy}{4\pi r^4} \Bigg[\frac{R_0^2 + z^2 - 2\tau R_0}{R_0} H(\tau - R_0)$$

$$- \alpha^{1/2}(1 + b^2) \frac{R_1^2 + z^2/\alpha}{R_1} H(\tau - R_1)$$

$$+ \left(2 + \frac{b^2 R_2^2 + b^2 z^2}{R_2} \right) H(\tau - R_2)$$

$$- \frac{(1 + b^2)^{1/2}}{2} \left\{ \frac{r^2 b^2 + 2m^2}{M} + \frac{r^2 b^2 + 2n^2}{N} \right\}$$

$$\times \{H(\tau - R_2) - H(\tau - R_1)\} \Bigg],$$

(3.10.56)

$$u_3(x,y,z,\tau) = \frac{xz\kappa}{4\pi r^2 |z| (\beta - \alpha)} \Bigg[\frac{z}{\alpha^{1/2} R_1} H(\tau - R_1) - \frac{|z|}{R_2} H(\tau - R_2)$$

$$+ \frac{1}{2} \left(\frac{n}{N} - \frac{m}{M} \right) \{H(\tau - R_2) - H(\tau - R_1)\} \Bigg],$$

(3.10.57)

and

$$w_3(x,y,z,\tau) = \frac{1}{4\pi} \Bigg[-\frac{b^2}{\alpha^{1/2} R_1} H(\tau - R_1) + \frac{1 + b^2}{R_2} H(\tau - R_2)$$

$$- \frac{b^2(1 + b^2)^{1/2}}{2} \left(\frac{1}{M} + \frac{1}{N} \right) \{H(\tau - R_2) - H(\tau - R_1)\} \Bigg].$$

(3.10.58)

Equations (3.10.55)–(3.10.58) together with (3.9.1) completely determine the nine components of the Green's tensor. In general there are three sheets of the wavefront, $\tau = R_2$ (spherical) $\tau = R_1$ (ellipsoid of revolution) and $\tau = R_0$ (ellipsoid of revolution). Each sheet carries a jump discontinuity, which may be smoothed out if the coefficient of the step function vanishes at the point in question on the front. (Incidentally, had the time dependence of the point load been $\delta(\tau)$ rather than $H(\tau)$, then the front would have carried a δ-function singularity.) The sheets of the front $\tau = R_0$ and $\tau = R_2$ touch on the z-axis (symmetry axis) at $z = \pm \tau$. Hence for this solid (and in general for any unconstrained transversely isotropic solid) multiple characteristics are always present in contrast to the isotropic situation. The wave front sheet $\tau = R_0$ may intersect (or touch) the front $\tau = R_1$ depending on the relation of the elastic parameters δ and β.

By taking a limit as $\tau \to \infty$ of equations (3.10.55)–(3.10.58) the components of the static Green's tensor can be obtained. These components are

$$u_1(x, y, z, \infty) = \frac{1}{4\pi r^4}\left\{\frac{r^4}{\delta R_0} - \frac{y^2 R_0^2 - x^2 z^2}{R_0} + \alpha^{1/2}(1 + b^2)\frac{y^2 R_1^2 + x^2 z^2/\alpha}{R_1}\right.$$

$$\left. - \frac{b^2 y^2 R_2^2 - b^2 x^2 z^2}{R_2}\right\}, \tag{3.10.59}$$

$$u_2(x, y, z, \infty) = \frac{xy}{4\pi r^4}\left\{\frac{R_0^2 + z^2}{R_0} - \alpha^{1/2}(1 + b^2)\frac{R_1^2 + z^2/\alpha}{R_1} + \frac{b^2 R_2^2 + b^2 z^2}{R_2}\right\}, \tag{3.10.60}$$

$$u_3(x, y, z, \infty) = \frac{xz\kappa}{4\pi r^2(\beta - \alpha)}\left\{\frac{1}{\alpha^{1/2} R_1} - \frac{1}{R_2}\right\}, \tag{3.10.61}$$

and

$$w_3(x, y, z, \infty) = \frac{1}{4\pi}\left\{-\frac{b^2}{\alpha^{1/2} R_1} + \frac{1 + b^2}{R_2}\right\}. \tag{3.10.62}$$

11. Transform Inversion for the General Case Along the Symmetry Axis

In the present section, the transformed displacements $\bar{u}_1^*$ and $\bar{w}_3^*$ recorded in equations (3.9.4) and (3.9.5) will be inverted at any arbitrary point z on the symmetry axis ($x = y = 0$). Unlike the previous section, where $\gamma = \alpha + \beta$, no special conditions need be imposed on the material constants α, β and γ. However, in order that the results be relevant to hexagonal crystals, it will be assumed (as in section 3.4) that

$$\alpha > 1 \quad \text{and} \quad \beta > 1. \tag{3.11.1}$$

By restricting the analysis to the symmetry axis of the solid, it is possible to obtain explicit closed-form expressions for the various displacement components. These results constitute the first (three-dimensional, time-dependent) extension of Stokes' [45] celebrated solution for an isotropic solid to a (physically realizable) anisotropic solid (Payton [49]). Cameron and Eason [44] studied the transverse plane motion ($z = 0$) for this problem, but the result is more complicated and requires numerical integration in order to obtain displacement values.

Chee-Seng [50], extending the work of Lighthill [15], has analyzed the axial wave motion in (not necessarily elastic) transversely isotropic media with particular emphasis on magnetohydrodynamics. Buchwald [51] has treated the steady-state version of the elastic wave problem and found an asymptotic solution valid for the far field.

The integration technique used to find the axial displacement w_3 will now be described. It will only be necessary to present the main features of the inversion, since the analysis has much in common with the two-dimensional inversion problem treated in sections 3.2 and 3.4.

Applying the inversion formula (3.9.3) to the expression (3.9.5) for $\bar{w}_3^*$ gives

$$w_3(x, y, z, \tau) = \frac{1}{2\pi i} \int_c e^{s\tau} ds \, \frac{1}{(2\pi)^3} \int_{-\infty}^{+\infty} \int_{-\infty}^{+\infty} \int_{-\infty}^{+\infty} \frac{-K(i\xi, i\eta, i\zeta, s)}{sQ(i\xi, i\eta, i\zeta, s)}$$

$$\times \, e^{i(\xi x + \eta y + \zeta z)} d\xi d\eta d\zeta. \tag{3.11.2}$$

Introduce spherical coordinates (R, θ, ϕ) by

$$\xi = R \sin \theta \cos \phi, \qquad \eta = R \sin \theta \sin \phi \quad \text{and} \quad \zeta = R \cos \theta \tag{3.11.3}$$

with $0 \leqslant \theta \leqslant \pi$ and $0 \leqslant \phi < 2\pi$. Then along the symmetry axis,

$$w_3(0, 0, z, \tau) = \text{Re} \, \frac{1}{2\pi i} \int_c e^{s\tau} ds \, \frac{1}{(2\pi)^2} \int_0^\infty R^2 \, dR$$

$$\times \int_0^\pi \frac{-K(iR \sin \theta, 0, iR \cos \theta, s)}{sQ(iR \sin \theta, 0, iR \cos \theta, s)} \, e^{iR|z| \cos \theta} \sin \theta d\theta, \tag{3.11.4}$$

having used the transverse isotropy properties of the K and Q operators to perform the ϕ integration. Next set $s = \sigma R$ and again use the homogeneous properties of the K and Q operators so that

$$w_3 = \text{Re} \, \frac{1}{2\pi i} \int_c e^{\sigma R \tau} d\sigma \, \frac{1}{(2\pi)^2} \int_0^\infty dR$$

$$\times \int_0^\pi \frac{-K(i \sin \theta, 0, i \cos \theta, \sigma)}{\sigma Q(i \sin \theta, 0, i \cos \theta, \sigma)} \, e^{iR|z| \cos \theta} \sin \theta d\theta. \tag{3.11.5}$$

At this stage the σ-integration can easily be performed, followed by the R-integration (after suitably interpreting the R-divergent integral as in equation (3.2.8)). Recombining the various terms in the integrand by partial fractions (in reverse) allows w_3 to be written as

$$w_3 = \text{Re} \, \lim_{\epsilon \to 0+} \frac{i}{(2\pi)^2 \tau} \int_0^\pi \sin \theta d\theta \, \frac{K(\sin \theta, 0, \cos \theta, g)}{gQ(\sin \theta, 0, \cos \theta, g)}, \tag{3.11.6}$$

where

$$g(\theta) = |\bar{z}| \cos \theta + i\epsilon, \tag{3.11.7}$$

and

$$\bar{z} = z/\tau. \tag{3.11.8}$$

The substitution

$$p = \cot \theta, \tag{3.11.9}$$

converts the integral (3.11.6) into

$$w_3 = \text{Re} \, \lim_{\epsilon \to 0+} \frac{i}{(2\pi)^2 \tau} \int_{-\infty}^{+\infty} \frac{K(1, 0, p, h)}{hQ(1, 0, p, h)} \, dp, \tag{3.11.10}$$

where

$$h = |\bar{z}|p + i\epsilon, \tag{3.11.11}$$

By closing the contour in the upper half p-plane, the integral (3.11.10) may be reduced to a residue calculation. However, some caution must be exercised in performing this calculation. The integrand of (3.11.10) contains two types of poles, those which remain in the upper half p-plane when $\epsilon = 0$ and those which drift down onto the real axis as $\epsilon \to 0+$. The contributions to the displacements w_3 and u_1 from the latter poles (i.e., real poles which are perturbed above the real axis when $\epsilon > 0$) are denoted by w_3^{**} and u_1^{**}. The residue evaluation (3.11.10) then gives

$$w_3(0,0,z,\tau) = \frac{1}{2\pi\tau} \operatorname{Re} \sum_p \left\{ \text{residues of } \frac{-K(1,0,p,|\bar{z}|p)}{|\bar{z}|pQ(1,0,p,|z|p)} \right.$$

$$\left. \text{in the upper half } p\text{-plane} \right\} + w_3^{**}. \tag{3.11.12}$$

Using a similar argument, the other three displacement components mentioned in section 3.9 and defined in equations (1.7.3) are given by

$$u_1(0,0,z,\tau) = \frac{1}{2\pi\tau} \operatorname{Re} \sum_p \left\{ \text{residues of } -\frac{Q(1,0,p,|\bar{z}|p) + M(1,0,p,|\bar{z}|p)/2}{|\bar{z}|pL(1,0,p,|\bar{z}|p)Q(1,0,p,|\bar{z}|p)} \right.$$

$$\left. \text{in the upper half } p\text{-plane} \right\} + u_1^{**}. \tag{3.11.13}$$

and

$$u_2(0,0,z,\tau) \equiv 0, \qquad u_3(0,0,z,\tau) \equiv 0. \tag{3.11.14}$$

In view of equations (3.11.14) and (3.9.1) it is seen that only the primary displacement components u_1, v_2 and w_3 of the Green's tensor differ from zero along the symmetry axis of a solid subjected to an arbitrarily oriented point load on this axis.

Complex Poles which Contribute to u_1 and w_3

From equations (1.5.20), (1.5.21) and (1.7.5) the p-dependence of the various terms in (3.11.12) and (3.11.13) may be written

$$K(1,0,p,|\bar{z}|p) = (1-\bar{z}^2)p^2 + \beta, \tag{3.11.15}$$

$$Q(1,0,p,|\bar{z}|p) = (\alpha - \bar{z}^2)(1-\bar{z}^2)p^4 + [\gamma - (\beta+1)\bar{z}^2]p^2 + \beta \equiv f(p), \tag{3.11.16}$$

$$M(1,0,p,|\bar{z}|p) = [(1+\alpha\delta - \gamma) - (\delta - \beta)\bar{z}^2]p^2 + (\delta - \beta), \tag{3.11.17}$$

and

$$L(1,0,p,|\bar{z}|p) = (1-\bar{z}^2)p^2 + \delta \equiv l(p). \tag{3.11.18}$$

Of interest for the w_3 and u_1 residue calculations are the upper half p-plane zeros of $l(p)$ and $f(p)$. An examination of (3.11.18) shows that for $0 \leqslant |\bar{z}| < 1$,

$$l(p) = 0 \quad \text{at} \quad p_3 = i[\delta/(1-\bar{z}^2)]^{1/2}, \tag{3.11.19}$$

otherwise $l(p)$ has no upper half p plane zeros. As for the complex zeros of $f(p)$, note that the expression for $f(p)$ in (3.11.16) has been previously encountered in equation (3.4.2) of section 3.4. There the relevant zeros of $f(p)$ were listed as a function of $\bar{z}$ according to the range of γ. There is no need to repeat these roots here. One modification of the $|\bar{z}|$ range in equation (3.4.13) is necessary in adapting those poles of the former 2-dimensional problem to the present 3-dimensional case. The replacement equation, in which $\bar{z}_1 = |\bar{z}|$ is eliminated from the $|\bar{z}|$ interval, now reads

$$\bar{z}_1 < |\bar{z}| < 1 \text{ no complex roots.} \tag{3.11.20}$$

Real Poles which Contribute to u_1^{**} and w_3^{**}

Suppose that q_0 is a real root of $Q(1, 0, p, |\bar{z}|p) = 0$. Then the corresponding root of $Q(1, 0, p, |\bar{z}|p + i\epsilon) = 0$ will be $p = q_0 + i\epsilon q_1 + \ldots$. The question is the sign of q_1, since only roots for which $q_1 > 0$ contribute to the residue calculation. By expanding Q in a Taylor series in ϵ, the value of q_1 is found to be $q_1 = -[Q_u/Q_v + |\bar{z}|]^{-1}$ where $Q_u = (\partial/\partial u)Q(1, 0, u, v)$, $Q_v = (\partial/\partial v)Q(1, 0, u, v)$ and u and v are to be replaced by q_0 and $|\bar{z}|q_0$ respectively, after differentiation.

The curve M: $Q(1, 0, u, v) = 0$ in the u, v plane has the slope $dv/du = -Q_u/Q_v$. Note that the substitution $p = u/v$, $q = 1/v$ maps the M-curve onto the normal curve N of section 2.2. In this way the M-curve geometric properties can be deduced. The sign of q_1 is determined by (slope of line $v = |\bar{z}|u$) − (slope of curve M). If this quantity is negative, then the corresponding p-pole (for $\epsilon > 0$) lies in the upper half p-plane and hence contributes to the residue calculation. Also, since Q_u/Q_v is even in q_0, if the real pole q_0 contributes to the residue then so does the $-q_0$. For this reason only the first quadrant of the u, v plane need be considered.

From the shape of the M-curve it is clear that the key question is the existence of a point on the curve where the (curve slope) = (slope of the line $v = |\bar{z}|u$). Such a point corresponds to a conical point on the symmetry axis of the wavefront (at $z = \pm \bar{z}_1 \tau$) and corresponds to a bitangent on the slowness curve. If there is no such point, there will be no residue contribution from the real poles. For the crystals to be treated in the next section, only apatite, cadmium and zinc have real poles (for $\epsilon = 0$) which are perturbed to the upper half p-plane when $\epsilon > 0$. The relevant p-poles are $p = p_5$ and $p = -p_5$ and fit into the scheme presented by equations (3.4.4)–(3.4.14) as follows

$$\gamma < (\beta + 1) \text{ and } (\gamma^2 - 4\alpha\beta) < 0$$

$$\bar{z}_1 < |\bar{z}| < 1 \qquad p = p_5 \equiv \left[\frac{-\{\gamma - (\beta + 1)\bar{z}^2\} + D^{1/2}}{2(\alpha - \bar{z}^2)(1 - \bar{z}^2)} \right]^{1/2}$$

$$\text{and} \quad p = -p_5. \tag{3.11.21}$$

Displacement Expressions for u_1 and w_3

The residue calculations necessary to complete the formulas (3.11.12) and (3.11.13) for the axial displacement components u_1 and w_3 can now be easily performed, using the p-plane pole locations as given above. After some simplification, the final results are given by

$(\alpha + \beta) < \gamma < (1 + \alpha\beta)$

$$0 \leqslant |\bar{z}| \leqslant 1, \qquad 4\pi\,|\bar{z}|\,w_3(0, 0, z, \tau) = 1, \qquad (3.11.22)$$

$$1 \leqslant |\bar{z}| < \sqrt{\alpha}, \qquad 4\pi\,|\bar{z}|\,w_3(0, 0, z, \tau) = h(\bar{z}), \qquad (3.11.23)$$

$$\sqrt{\alpha} \leqslant |\bar{z}| < \infty, \qquad 4\pi\,|\bar{z}|\,w_3(0, 0, z, \tau) = 0, \qquad (3.11.24)$$

and

$$0 \leqslant |\bar{z}| < 1, \qquad 8\pi\,|\bar{z}|\,u_1(0, 0, z, \tau) = \frac{1}{\delta} + \frac{1}{\beta}, \qquad (3.11.25)$$

$$1 < |\bar{z}| \leqslant \sqrt{\alpha}, \qquad 8\pi\,|\bar{z}|\,u_1(0, 0, z, \tau) = k(\bar{z}), \qquad (3.11.26)$$

$$\sqrt{\alpha} \leqslant |\bar{z}| < \infty, \qquad 8\pi\,|\bar{z}|\,u_1(0, 0, z, \tau) = 0. \qquad (3.11.27)$$

$(\beta + 1) < \gamma < (\alpha + \beta)$ and $(\gamma^2 - 4\alpha\beta) < 0$

Here the results are the same as those set down in equations (3.11.22)–(3.11.27).

$\gamma < (\beta + 1)$ and $(\gamma^2 - 4\alpha\beta) < 0$

$$0 \leqslant |\bar{z}| < \bar{z}_1, \qquad 4\pi\,|z|\,w_3(0, 0, z, \tau) = 1, \qquad (3.11.28)$$

$$\bar{z}_1 < |\bar{z}| \leqslant 1, \qquad 4\pi\,|z|\,w_3(0, 0, z, \tau) = 2h(\bar{z}), \qquad (3.11.29)$$

$$1 \leqslant |\bar{z}| < \sqrt{\alpha}, \qquad 4\pi\,|z|\,w_3(0, 0, z, \tau) = h(\bar{z}), \qquad (3.11.30)$$

$$\sqrt{\alpha} \leqslant |\bar{z}| < \infty, \qquad 4\pi\,|z|\,w_3(0, 0, z, \tau) = 0, \qquad (3.11.31)$$

and

$$0 \leqslant |\bar{z}| < \bar{z}_1, \qquad 8\pi\,|z|\,u_1(0, 0, z, \tau) = \frac{1}{\delta} + \frac{1}{\beta}, \qquad (3.11.32)$$

$$\bar{z}_1 < |\bar{z}| < \bar{z}_1, \qquad 8\pi\,|z|\,u_1(0, 0, z, \tau) = \frac{1}{\delta} + 2k(\bar{z}), \qquad (3.11.33)$$

$$1 < |\bar{z}| \leqslant \sqrt{\alpha}, \qquad 8\pi\,|z|\,u_1(0, 0, z, \tau) = k(\bar{z}), \qquad (3.11.34)$$

$$\sqrt{\alpha} \leqslant |\bar{z}| < \infty, \qquad 8\pi\,|z|\,u_1(0, 0, z, \tau) = 0. \qquad (3.11.35)$$

The two functions $h(\bar{z})$ and $k(\bar{z})$ introduced into the displacement expressions above are defined by

$$h(\bar{z}) = \frac{1}{2} - \frac{2(1 - \bar{z}^2) - \{\gamma - (\beta + 1)\bar{z}^2\}}{2D^{1/2}} \qquad (3.11.36)$$

and

$$k(\bar{z}) = \frac{1}{2\beta} + \frac{(\beta - 1)\bar{z}^2 + (\gamma - 2\alpha\beta)}{2\beta D^{1/2}}, \qquad (3.11.37)$$

where $D(\bar{z})$ was introduced in (3.4.15). Both the functions h and k are defined on the interval $1 < |\bar{z}| < \sqrt{\alpha}$ and have the limit values

$$\lim_{|\bar{z}| \to 1+0} h(\bar{z}) = 1 \quad \text{for} \quad \gamma > (\beta + 1)$$
$$= 0 \quad \text{for} \quad \gamma < (\beta + 1), \tag{3.11.38}$$

$$\lim_{|\bar{z}| \to 1+0} k(\bar{z}) = -\frac{1 + \alpha\beta - \gamma}{\beta(\gamma - \beta - 1)} \quad \text{for} \quad \gamma > (\beta + 1)$$

$$= -\frac{\alpha - 1}{\beta + 1 - \gamma} \quad \text{for} \quad \gamma < (\beta + 1), \tag{3.11.39}$$

$$\lim_{|\bar{z}| \to \sqrt{\alpha} - 0} h(\bar{z}) = \frac{\alpha - 1}{(\beta + 1)\alpha - \gamma}, \tag{3.11.40}$$

$$\lim_{|\bar{z}| \to \sqrt{\alpha} - 0} k(\bar{z}) = 0. \tag{3.11.41}$$

The w_3 displacement component experiences jump discontinuities at $|\bar{z}| = \sqrt{\alpha}$ and at $|\bar{z}| = \bar{z}_1$. Both displacement components are singular at $|\bar{z}| = \bar{z}_1$ when $\gamma < (\beta + 1)$ and $(\gamma^2 - 4\alpha\beta) < 0$. This reciprocal square root singularity (due to the vanishing of $D(\bar{z}_1)$) is in agreement with Buchwald [51]. After passage of the trailing sheet of the

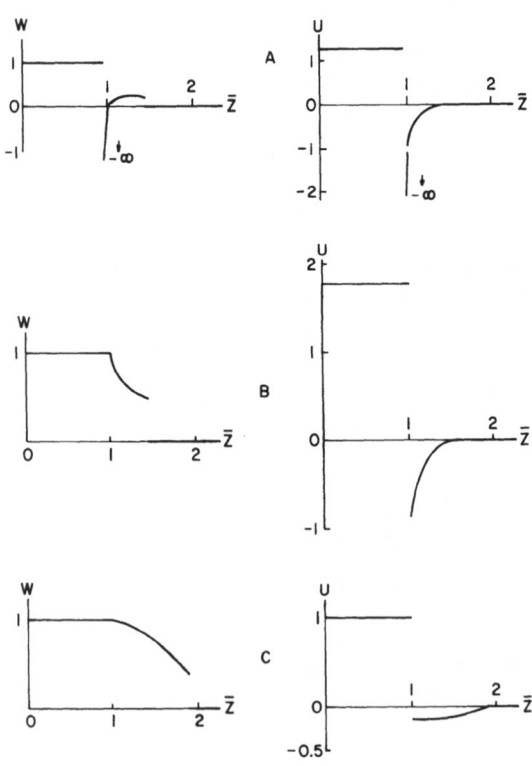

Figure 57.   Normalized Displacement Components w and u versus $\bar{z}$ for: A. Apatite B. Beryllium and C. Beryl.

wavefront, $|z| = \tau$ or $|z| = \bar{z}_1 \tau$ as the case may be, the displacement components immediately assume their static (time-independent) values. This deformation without restriction to the symmetry axis, can of course be verified independently of the above analysis.

12. Application to Some Hexagonal Crystals

While the above analysis of the roots of $f(p) = 0$ and the subsequent residue calculations for u_1 and w_3 was not exhaustive it does include all the 13 crystals listed in tables 7 and 11. One notable omission is the case $\gamma = (\alpha + \beta)$, but this case has been treated separately in section 3.10.

Figures 57–60 show the variation of the normalized displacement components $4\pi |z| w_3(0, 0, z, \tau) \equiv w$ and $8\pi |z| u_1(0, 0, z, \tau) \equiv u$ as a function of the single variable $\bar{z}$ for the crystals listed in table 7. The main features of these graphs have been anticipated in the discussion at the end of the last section. In particular apatite, cadmium and zinc have singularities at the conical point on the wavefront. The respective values of $\bar{z}_1$ for these materials are 0.9633, 0.9997 and 0.8871. Cadmium

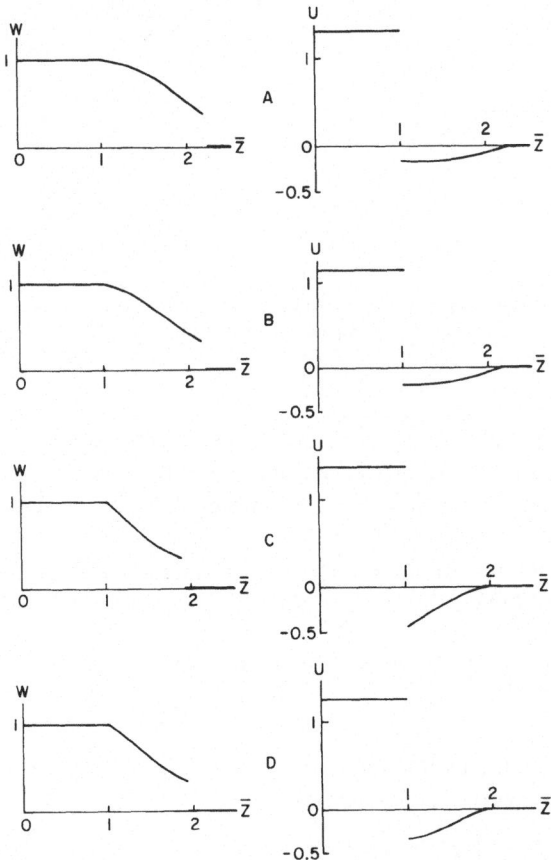

Figure 58. Normalized Displacement Components w and u versus $\bar{z}$ for A. Cobalt, B. Ice, C. Hafnium and D. Magnesium.

111

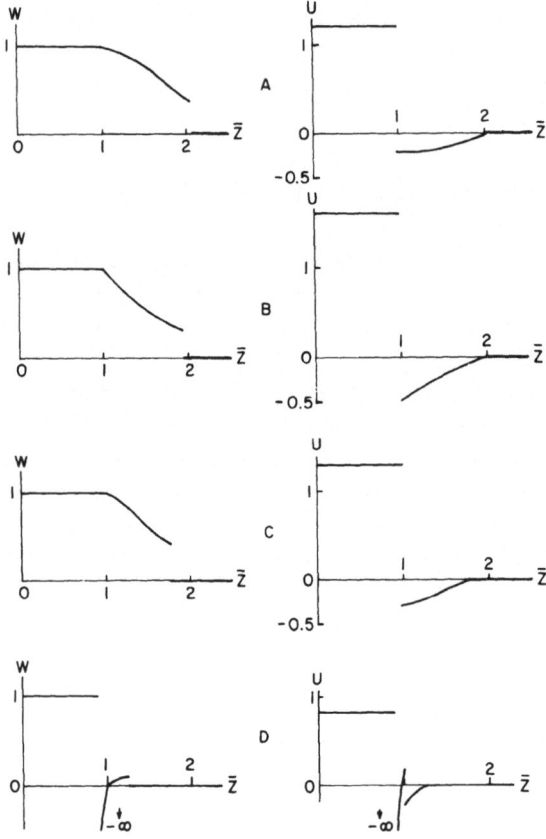

Figure 59.   Normalized Displacement Components w and u versus $\bar{z}$ for: A. Rhenium B. Titanium
C. Yttrium and D. Zinc.

is an interesting special case. Since γ and $(\beta + 1)$ are very close in numerical value, two inflection points on the normal curve almost coalesce on the symmetry axis. This results in a large displacement as $\bar{z} \to 1 + 0$ which is extremely close to the singularity at $\bar{z} = \bar{z}_1$.

It is instructive to search for possible maximum and minimum values of the functions $h(\bar{z})$ and $k(\bar{z})$ defined by equations (3.11.36) and (3.11.37). Now

$$dh/d\bar{z} = 2(\alpha\beta + 1 - \gamma)\frac{\bar{z}[(\gamma - 2\beta) + (\beta - 1)\bar{z}^2]}{D^{3/2}}. \qquad (3.12.1)$$

Under suitable conditions this slope will change sign in the interval $1 < \bar{z} < \sqrt{\alpha}$ with the result that h takes on a maximum value

$$h_{max} = \frac{1}{2}\left[1 - \left\{\frac{(\beta + 1 - \gamma)(\beta - 1)}{(\beta + \alpha - \gamma)\beta}\right\}^{1/2}\right], \qquad (3.12.2)$$

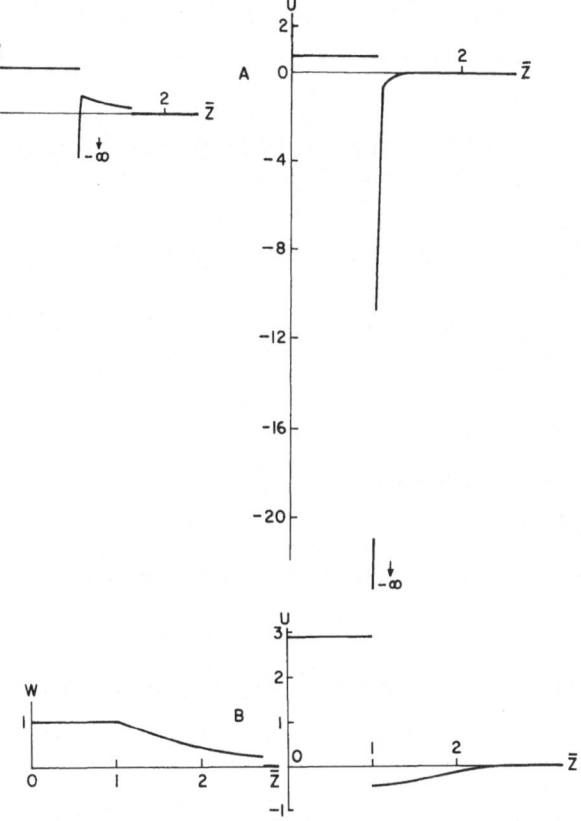

Figure 60. Normalized Displacement Components w and u versus $\bar{z}$ for A. Cadmium and B. Thallium.

at

$$|\bar{z}| = \left[\frac{2\beta - \gamma}{\beta - 1}\right]^{1/2}. \tag{3.12.3}$$

This is the situation for apatite and cadmium; for the other eleven crystals $h(\bar{z})$ is monotonic. Similarly

$$dk/d\bar{z} = -2(\alpha\beta + 1 - \gamma)\frac{\bar{z}[(\gamma - 2\alpha) - (\beta - 1)\bar{z}^2]}{D^{3/2}}, \tag{3.12.4}$$

which under suitable conditions will give rise to a minimum value

$$k_{\min} = \frac{1}{2\beta}\left[1 - \left\{\frac{(\alpha\beta + \alpha - \gamma)(\beta - 1)}{\gamma - \alpha - \beta}\right\}^{1/2}\right], \tag{3.12.5}$$

at

$$|\bar{z}| = \left[\frac{\gamma - 2\alpha}{\beta - 1}\right]^{1/2}. \tag{3.12.6}$$

113

This is the situation for beryl, cobalt, ice and rhenium; for the other nine crystals $k(\bar{z})$ is monotonic.

13. Herglotz–Petrowski Formula for the Displacement Field as an Integral Over the Slowness Surface

In the integral representation of w_3 given in equation (3.11.2) make the change of (integration) variables

$$\xi = r\omega_1, \qquad \eta = r\omega_2 \quad \text{and} \quad \zeta = r\omega_3 \tag{3.13.1}$$

where

$$\omega_1^2 + \omega_2^2 + \omega_3^2 = 1.$$

Also, the volume element becomes

$$d\xi d\eta d\zeta = r^2 dr d\omega, \tag{3.13.2}$$

where $d\omega$ = surface area element on the unit sphere $|\hat{\omega}| = 1$ with

$$\hat{\omega} = (\omega_1, \omega_2, \omega_3). \tag{3.13.3}$$

Denote the position vector (which remains fixed throughout the integration) by

$$\mathbf{x} = (x, y, z). \tag{3.13.4}$$

In a shortened notation

$$K(i\xi, i\eta, i\zeta, s) \equiv K(ir\hat{\omega}, s),$$

and

$$Q(i\xi, i\eta, i\zeta, s) \equiv Q(ir\hat{\omega}, s). \tag{3.13.5}$$

Next stretch the Laplace variable s to σ by

$$s = \sigma r, \tag{3.13.6}$$

and note that

$$K(ir\hat{\omega}, \sigma r) = r^2 K(i\hat{\omega}, \sigma) \quad \text{since } K \text{ is homogeneous of degree 2},$$

$$Q(ir\hat{\omega}, \sigma r) = r^4 Q(i\hat{\omega}, \sigma) \quad \text{since } Q \text{ is homogeneous of degree 4}. \tag{3.13.7}$$

Inserting the above changes into the multiple integral (3.11.2) gives

$$w_3 = \text{Re} \frac{1}{2\pi i} \int_c e^{r\sigma\tau} d\sigma \frac{1}{(2\pi)^3} \int_{|\hat{\omega}|=1} d\omega \int_0^\infty e^{ir(\mathbf{x} \cdot \hat{\omega})} dr \frac{-K(i\hat{\omega}, \sigma)}{\sigma Q(i\hat{\omega}, \sigma)}. \tag{3.13.8}$$

Here Re = real part, may be inserted without loss in generality due to the properties of the integrand. First perform the σ-integration. Under the assumption of strict hyperbolicity for the Q operator (which is certainly valid for the crystals listed in table 7), the complex σ-plane will contain one real pole at the origin and four pure imaginary poles

$$\sigma_j = iv_j(\hat{\omega}) \qquad j = 1, 2, 3, 4 \tag{3.13.9}$$

where $v_j(\hat{\omega})$ is a root of $Q(\hat{\omega}, v)$. Since Q is a biquadratic in v, it is permissible to take the roots as

$$v_1(\hat{\omega}) = -v_2(\hat{\omega}), \qquad v_3(\hat{\omega}) = -v_4(\hat{\omega}) \tag{3.13.10}$$

Also, because Q is an even function of ω_1, ω_2 and w_3, it follows that

$$v(\hat{\omega}) = v(-\hat{\omega}) \tag{3.13.11}$$

i.e. the plane wave speed in the direction $\hat{\omega}$ is the same as the plane wave speed in the direction $-\hat{\omega}$.

By a residue calculation, the σ-integration gives

$$w_3 = \mathrm{Re}\, \frac{1}{(2\pi)^3} \int_{|\hat{\omega}|=1} \frac{K(\hat{\omega}, 0)}{Q(\hat{\omega}, 0)} \, d\omega \int_0^\infty e^{ir(\mathbf{x}\cdot\hat{\omega})} \, dr$$

$$+ \mathrm{Re}\, \sum_{j=1}^4 \frac{1}{(2\pi)^3} \int_{|\hat{\omega}|=1} \frac{K(\hat{\omega}, v_j(\hat{\omega})) \, d\omega}{v_j(\hat{\omega}) Q_\sigma(\hat{\omega}, v_j(\hat{\omega}))} \int_0^\infty e^{ir(v_j(\hat{\omega})\tau + \mathbf{x}\cdot\hat{\omega})} \, dr, \tag{3.13.12}$$

having used the appropriate homogeneous property of K, Q and Q_σ. As in section 3.2, the divergent integral

$$\int_0^\infty e^{ir\lambda} \, dr \quad \text{with} \quad \mathrm{Im}\, \lambda = 0$$

is interpreted as

$$\lim_{\epsilon \to 0+} \int_0^\infty e^{ir\lambda - \epsilon r} \, dr$$

so that from (3.2.9)

$$\mathrm{Re} \int_0^\infty e^{ir\lambda} \, dr = \pi \delta(\lambda). \tag{3.13.13}$$

Thus

$$w_3 = \frac{1}{8\pi^2} \int_{|\hat{\omega}|=1} \frac{K(\hat{\omega}, 0)}{Q(\hat{\omega}, 0)} \, \delta(\mathbf{x}\cdot\hat{\omega}) \, d\omega$$

$$+ \sum_{j=1}^4 \frac{1}{8\pi^2} \int_{|\hat{\omega}|=1} \frac{K(\hat{\omega}, v_j(\hat{\omega}))}{v_j(\hat{\omega}) Q_\sigma(\hat{\omega}, v_j(\hat{\omega}))} \, \delta(v_j(\hat{\omega})\tau + \mathbf{x}\cdot\hat{\omega}) \, d\omega. \tag{3.13.14}$$

An examination of (3.11.2) reveals that w_3 is an even function of x, y and z so that the vector $\mathbf{x}$ may be replaced by $-\mathbf{x}$ in (3.13.14) if convenient. This fact, together with (3.13.10) enables (3.13.14) to be rewritten as

$$w_3 = \frac{1}{8\pi^2} \int_{|\hat{\omega}|=1} \frac{K(\hat{\omega}, 0)}{Q(\hat{\omega}, 0)} \, \delta(\mathbf{x}\cdot\hat{\omega}) \, d\omega$$

$$+ \sum_{j=2,4} \frac{1}{4\pi^2} \int_{|\hat{\omega}|=1} \frac{K(\hat{\omega}, v_j(\hat{\omega}))}{v_j(\hat{\omega})Q_\sigma(\hat{\omega}, v_j(\hat{\omega}))} \, \delta(v_j(\hat{\omega})\tau - \mathbf{x} \cdot \hat{\omega}) \, d\omega. \quad (3.13.15)$$

Fix $\mathbf{x}$ and allow $\tau \to +\infty$. Then since $v_j(\hat{\omega}) > 0$ it follows that $(v_j(\hat{\omega})\tau - \mathbf{x} \cdot \hat{\omega}) > 0$ and hence, $\delta(v_j(\hat{\omega})\tau - \mathbf{x} \cdot \hat{\omega}) = 0$ for all unit vectors $\hat{\omega}$. Thus the first integral in the above expression evidently represents the static solution for the axial displacement. Denote this integral by w_{3s} so that

$$w_{3s} = \frac{1}{8\pi^2} \int_{|\hat{\omega}|=1} \frac{K(\hat{\omega}, 0)}{Q(\hat{\omega}, 0)} \, \delta(\mathbf{x} \cdot \hat{\omega}) \, d\omega. \quad (3.13.16)$$

The remaining two integrals in (3.13.15) will now be converted from integration over the unit sphere to integration over the slowness surface N of section 2.16. Toward this goal, consider the substitution

$$\mathbf{q} = \frac{\hat{\omega}}{v(\hat{\omega})}, \quad (3.13.17)$$

where $v = v_2$ or v_4 and of course $\mathbf{q} = \mathbf{q}_2$ or $\mathbf{q}_4$. Now since

$$Q(\hat{\omega}, v_j(\hat{\omega})) = 0 \quad \text{or} \quad v_j^4(\hat{\omega})Q(\mathbf{q}_j, 1) = 0 \quad (3.13.18)$$

and

$$Q(\mathbf{q}, 1) = Q(\mathbf{q}, -1),$$

it follows from (2.16.2) that

$$Q(\mathbf{q}_j, 1) = 0$$

implies that the vector $\mathbf{q}_j$ lies on the jth sheet of the slowness surface N. Now if $\mathbf{q}$ satisfies

$$Q(\mathbf{q}, 1) = 0, \quad (3.13.19)$$

and $v(\hat{\omega})$ satisfies

$$Q(\hat{\omega}, v(\hat{\omega})) = 0, \quad (3.13.20)$$

then replacing $\hat{\omega}$ by $\mathbf{q}$ and $v(\hat{\omega})$ by $v(\mathbf{q})$ in (3.13.20) gives

$$Q(\mathbf{q}, v(\mathbf{q})) = 0 \quad (3.13.21)$$

which, after comparing with (3.13.19) shows that

$$v(\mathbf{q}) = 1 \quad \text{for} \quad \mathbf{q} \text{ on } N. \quad (3.13.22)$$

A unit normal vector to the jth sheet of N is thus given by

$$\hat{n}_j = \frac{\nabla v(\mathbf{q}_j)}{|\nabla v(\mathbf{q}_j)|}. \quad (3.13.23)$$

If μ_j is the angle between the vectors $\hat{n}_j$ and $\mathbf{q}_j$ on N_j then

$$\cos \mu_j = \frac{\mathbf{q}_j}{|\mathbf{q}_j|} \cdot \hat{n}_j. \quad (3.13.24)$$

Now let the surface element $d\omega$ on the unit sphere project the area element dS_j on the sheet N_j. Then $d\omega$ and dS_j are related by spherical projection

$$d\omega = |\cos \mu_j| \frac{dS_j}{|q_j|^2} = \frac{|\hat{n}_j \cdot q_j| dS_j}{|q_j|^3} \tag{3.13.25}$$

or, from (3.13.23)

$$d\omega = \frac{|q_j \cdot \nabla v(q_j)|}{|q_j|^3 |\nabla v(\bar{q}_j)|} dS_i. \tag{3.13.26}$$

Next observe that since Q is homogeneous of degree 4, $Q(\lambda x_1, \lambda x_2, \lambda x_3, \lambda v) = \lambda^4 Q(x_1, x_2, x_3, v)$. Take $(\partial/\partial \lambda)|_{\lambda=1}$ so that

$$x_1 Q_1 + x_2 Q_2 + x_3 Q_3 + v Q_v = 4Q \tag{3.13.27}$$

Then if v is such that $Q(x_1, x_2, x_3, v) = 0$ it follows from (3.13.27) that

$$x \cdot \nabla_x Q + v Q_v = 0. \tag{3.13.28}$$

Next take $\partial/\partial x_1$ of

$$Q(x_1, x_2, x_3, v(x_1, x_2, x_3)) = 0 \tag{3.13.29}$$

so that

$$Q_1 + Q_v v_1 = 0, \tag{3.13.30}$$

hence

$$x \cdot \nabla_x Q + (x \cdot \nabla_x v) Q_v = 0. \tag{3.13.31}$$

Comparing (3.13.31) with (3.13.28) shows that, since $Q_v \neq 0$, (and in an obvious change in notation)

$$q \cdot \nabla_q v(q) = v(q) = 1 \tag{3.13.32}$$

for q on N by (3.13.22).

All of this allows the expression (3.13.26) to be simplified to

$$d\omega = \frac{dS_j}{|q_j|^3 |\nabla v(q_j)|}. \tag{3.13.33}$$

Now use

$$K(\hat{\omega}, v_j(\hat{\omega})) = v_j^2(\hat{\omega}) K(q_j, 1) \tag{3.13.34}$$

$$v_j(\hat{\omega}) Q_o(\hat{\omega}, v_j(\hat{\omega})) = v_j^4(\hat{\omega}) Q_o(q_j, 1) \tag{3.13.35}$$

and

$$\delta(v_j(\hat{\omega})\tau - x \cdot \hat{\omega}) = \frac{1}{v_j(\hat{\omega})} \delta(\tau - x \cdot q_j) \tag{3.13.36}$$

together with (3.13.33) to rewrite the integral expression for w_3 as

$$w_3 = w_{3s} + \sum_{j=2,4} \frac{1}{4\pi^2} \int_{N_j} \frac{K(q_j, 1)\delta(\tau - x \cdot q_j) dS_j}{Q_o(q_j, 1) |\nabla v(\bar{q}_j)|}, \tag{3.13.37}$$

having used $v_j^3(\hat{\omega})|q_j|^3 = |\hat{\omega}|^3 = 1$ from (3.13.17). In (3.13.37) the pair of integrals over the domain N_j for $j = 2$ and 4 may be (in a compressed notation) replaced by a single integral over the entire slowness surface, thus

$$w_3 = w_{3s} + \frac{1}{4\pi^2} \int_N \frac{K(\mathbf{q}, 1)\delta(\tau - \mathbf{x} \cdot \mathbf{q})\,dS}{Q_\sigma(\mathbf{q}, 1)|\nabla v(\mathbf{q})|}. \tag{3.13.38}$$

From equation (3.13.30), it follows by addition that

$$\nabla Q(\mathbf{q}, 1) + (\nabla v(\mathbf{q}))Q_v(\mathbf{q}, 1) = 0, \tag{3.13.39}$$

hence

$$|\nabla v(\mathbf{q})| = \frac{|\nabla Q(\mathbf{q}, 1)|}{|Q_v(\mathbf{q}, 1)|}. \tag{3.13.40}$$

Thus the expression (3.13.38) for w_3 becomes

$$w_3 = w_{3s} + \frac{1}{4\pi^2} \int_N \frac{K(\mathbf{q}, 1)\delta(\tau - \mathbf{x} \cdot \mathbf{q})\,dS}{(\text{sgn } Q_v(\mathbf{q}, 1))|\nabla Q(\mathbf{q}, 1)|} \tag{3.13.41}$$

where

$$Q_v(\bar{\mathbf{q}}, 1) = \left.\frac{\partial Q(\bar{\mathbf{q}}, v)}{\partial v}\right|_{v=1} \tag{3.13.42}$$

and

$$\nabla Q(\mathbf{q}, 1) = (\partial Q(q_1, q_2, p, 1)/\partial q_1, \partial Q(q_1, q_2, p, 1)/\partial q_2, \partial Q(q_1, q_2, p, 1)/\partial p). \tag{3.13.43}$$

The time dependent integral in equation (3.13.41), in which the integration domain covers the slowness surface N, is known as a Herglotz–Petrowski formula [52, 32]. The derivation given above follows that of Burridge [53]. A clear presentation is also given in Gel'fand and Shilov [8]. The Herglotz–Petrowski formula is useful in analyzing the displacement behaviour near the wavefront W and this will form the subject of section 3.15 below. Of course the other components of the Green's tensor may be placed in the Herglotz–Petrowski form, but that will not be done here, since the form for w_3 will be sufficient to illustrate the novelty of integration on the slowness surface.

Finally, before closing this section, a definitive statement will be made regarding the expression sgn $Q_v(\mathbf{q}, 1)$ which appears in (3.13.41). Taking $\mathbf{q} = (q_1, q_2, p)$ then from (1.5.21)

$$Q(\mathbf{q}, v) = \alpha p^4 + \gamma p^2(q_1^2 + q_2^2) + \beta(q_1^2 + q_2^2)^2 - (\alpha + 1)v^2 p^2$$
$$- (\beta + 1)v^2(q_1^2 + q_2^2) + v^4, \tag{3.13.44}$$

$$\partial Q/\partial v = 4v^3 - 2(\alpha + 1)vp^2 - 2(\beta + 1)v(q_1^2 + q_2^2), \tag{3.13.45}$$

$$\partial Q/\partial v|_{v=1} = 4 - 2(\alpha + 1)p^2 - 2(\beta + 1)(q_1^2 + q_2^2). \tag{3.13.46}$$

But from (2.16.3) $q_1^2 + q_2^2 = R^2 \sin^2\theta, p^2 = R^2 \cos^2\theta$ so that

$$\partial Q/\partial v|_{v=1} = 4 - 2R^2 B = 2A\left(\frac{1}{A} - R^4\right) \tag{3.13.47}$$

having used (1.5.27) and (2.16.4). Note here that A, B and R depend on θ. From (2.16.4) $R(\theta)$ satisfies

$$R^4 - \frac{B}{A}R^2 + \frac{1}{A} = 0 \tag{3.13.48}$$

so that, using the notation of (2.16.10) for the R-roots of (3.13.48)

$$R_+^2 R_-^2 = \frac{1}{A} \quad \text{and since} \quad R_+ > R_-$$

$$R_-^2 < \frac{1}{\sqrt{A}} \quad \text{and} \quad R_+^2 > \frac{1}{\sqrt{A}}. \tag{3.13.48}$$

Consequently on N_+

$$\left(\frac{1}{A} - R_+^4\right) < 0, \tag{3.13.49}$$

while on N_-

$$\left(\frac{1}{A} - R_-^4\right) > 0, \tag{3.13.50}$$

Thus, from (3.13.47), since $A(\theta) > 0$ from section 1.5,

$$\text{sgn } Q_v(\mathbf{q}, 1) = \begin{cases} +1 & \text{on} \quad N_- \\ -1 & \text{on} \quad N_+ \end{cases} \tag{3.13.51}$$

This last result allows the slowness surface integral (3.13.41) to be written

$$w_3 = w_{3s} + \frac{1}{4\pi^2}\int_{N_-} \frac{K(\mathbf{q}_-, 1)\delta(\tau - \mathbf{x} \cdot \mathbf{q}_-)dS_-}{|\nabla Q(\mathbf{q}_-, 1)|}$$

$$- \frac{1}{4\pi^2}\int_{N_+} \frac{K(\mathbf{q}_+, 1)\delta(\tau - \mathbf{x} \cdot \mathbf{q}_+)dS_+}{|\nabla Q(\mathbf{q}_+, 1)|}. \tag{3.13.52}$$

14. Differential Geometry of the Slowness Surface N

With a view toward analyzing formula (3.13.52) further, certain differential geometry results will be derived in this section. Let

$$\mathbf{q}(\theta, \phi) = (R \sin\theta \cos\phi, R \sin\theta \sin\phi, R \cos\theta) \tag{3.14.1}$$

be the position vector in (q_1, q_2, p) space of a point on the slowness (or normal) surface N. In this θ, ϕ parametric representation of N, the θ and ϕ parameters range over the domain

$$0 \leqslant \theta \leqslant \pi \quad \text{and} \quad 0 \leqslant \phi < 2\pi, \tag{3.14.2}$$

as previously indicated in (2.16.3).

In equation (3.14.1)

$$R = R(\theta) = R_{\pm}(\theta) \tag{3.14.3}$$

depending upon which sheet of N is under discussion. Formulas for $R_+(\theta)$ and $R_-(\theta)$ can be found in (2.16.10). For the present, note that (when the conditions of strict hyperbolicity are satisfied)

$$R_+(\theta) > R_-(\theta) > 0 \tag{3.14.4}$$

Now

$$\mathbf{q}_\theta = R'\hat{e}_1 + R\hat{e}_2, \tag{3.14.5}$$

$$\mathbf{q}_\phi = R \sin \theta \, \hat{e}_3, \tag{3.14.6}$$

where

$$\hat{e}_1 = (\sin \theta \cos \phi, \sin \theta \sin \phi, \cos \theta), \tag{3.14.7}$$

$$\hat{e}_2 = (\cos \theta \cos \phi, \cos \theta \sin \phi, -\sin \theta), \tag{3.14.8}$$

and

$$\hat{e}_3 = (-\sin \phi, \cos \phi, 0). \tag{3.14.9}$$

The vectors $\hat{e}_1$, $\hat{e}_2$ and $\hat{e}_3$ form a righthanded orthonormal triad of base vectors (see figure 61). The surface normal vector $\mathbf{N}$ is found by

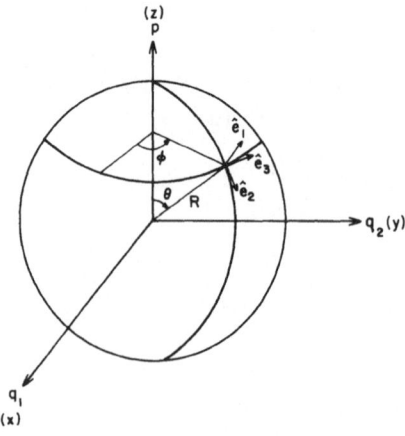

Figure 61.   Unit Vectors Used in Describing the Slowness Surface. Note that the p-axis is the symmetry axis.

$$\mathbf{N} = \mathbf{q}_\theta \times \mathbf{q}_\phi = R \sin\theta \, (R\hat{e}_1 - R'\hat{e}_2). \tag{3.14.10}$$

The fact that $\mathbf{N}$ vanishes when $\sin\theta = 0$ (i.e., at $\theta = 0$ and π) is not due to a singularity of the surface, but rather to a singularity of the parametric coordinate system (Stoker [40]). A unit vector normal to the slowness surface is found from (3.14.10) by

$$\hat{n} = \frac{\mathbf{N}}{|\mathbf{N}|} = \frac{R\hat{e}_1 - R'\hat{e}_2}{\sqrt{R^2 + R'^2}}. \tag{3.14.11}$$

Note that

$$\mathbf{q}_\theta \cdot \hat{n} = 0 \quad \text{and} \quad \mathbf{q}_\phi \cdot \hat{n} = 0 \tag{3.14.12}$$

since the coordinate vectors $\mathbf{q}_\theta$ and $\mathbf{q}_\phi$ lie in the tangent plane.

For use below, the following derivative formulas of the base vectors are useful,

$$\frac{\partial\hat{e}_1}{\partial\theta} = \hat{e}_2, \qquad \frac{\partial\hat{e}_2}{\partial\theta} = -\hat{e}_1, \qquad \frac{\partial\hat{e}_3}{\partial\theta} = 0, \tag{3.14.13}$$

$$\frac{\partial\hat{e}_1}{\partial\phi} = \sin\theta \, \hat{e}_3, \qquad \frac{\partial\hat{e}_2}{\partial\phi} = \cos\theta \, \hat{e}_3, \qquad \frac{\partial\hat{e}_3}{\partial\phi} = -\sin\theta \, \hat{e}_1 - \cos\theta \, \hat{e}_2. \tag{3.14.14}$$

By differentiation of equations (3.14.5) and (3.14.6) it is easily shown that

$$\mathbf{q}_{\theta\theta} = (R'' - R)\hat{e}_1 + 2R'\hat{e}_2, \tag{3.14.15}$$

and

$$\mathbf{q}_{\theta\theta} \cdot \hat{n} = -\frac{R^2 + 2R'^2 - RR''}{R^3} \frac{R^3}{\sqrt{R^2 + R'^2}}, \tag{3.14.16}$$

or

$$\mathbf{q}_{\theta\theta} \cdot \hat{n} = -(V + V'')\frac{R^3}{\sqrt{R^2 + R'^2}} \tag{3.14.17}$$

where

$$V(\theta) = \frac{1}{R(\theta)}. \tag{3.14.18}$$

Also

$$\mathbf{q}_{\theta\phi} = (R\cos\theta + R'\sin\theta)\hat{e}_3, \tag{3.14.19}$$

so

$$\mathbf{q}_{\theta\phi} \cdot \hat{n} = 0, \tag{3.14.20}$$

and

$$\mathbf{q}_{\phi\phi} = -R\sin\theta \, (\sin\theta \, \hat{e}_1 + \cos\theta \, \hat{e}_2), \tag{3.14.21}$$

so that

$$q_{\phi\phi} \cdot \hat{n} = \frac{R \sin \theta}{\sqrt{R^2 + R'^2}} (-R \sin \theta + R' \cos \theta),$$

$$= \frac{R \sin \theta}{\sqrt{R^2 + R'^2}} (R \cos \theta)'. \tag{3.14.22}$$

Incorporating the above results into the language of differential geometry (Stoker [40]) gives

First Fundamantal Form

$$I = dq \cdot dq = q_\theta \cdot q_\theta (d\theta)^2 + 2q_\theta \cdot q_\phi d\theta d\phi + q_\phi \cdot q_\phi (d\phi)^2$$
$$\equiv e(d\theta)^2 + 2f d\theta d\phi + g(d\phi)^2, \tag{3.14.23}$$

so that, from above

$$e = R^2 + R'^2, \qquad f = 0 \quad \text{and} \quad g = R^2 \sin^2\theta. \tag{3.14.24}$$

The surface area element on N is given by

$$dS = |q_\theta \times q_\phi| d\theta d\phi = \sqrt{R^2 + R'^2} R \sin \theta \, d\theta d\phi$$

$$= \sqrt{eg} \, d\theta d\phi. \tag{3.14.25}$$

Second Fundamental Form

$$II = q_{\theta\theta} \cdot \hat{n}\theta^2 + 2q_{\theta\phi} \cdot \hat{n}\theta\phi + q_{\phi\phi} \cdot \hat{n}\phi^2$$
$$\equiv l\theta^2 + 2m\theta\phi + n\phi^2, \tag{3.14.26}$$

so that, from the calculations of this section

$$l = -(V + V'') \frac{R^3}{\sqrt{R^2 + R'^2}}, \qquad m = 0 \tag{3.14.27}$$

and

$$n = \frac{R \sin \theta}{\sqrt{R^2 + R'^2}} (R \cos \theta)'. \tag{3.14.28}$$

The signs of l and n are of vital importance in establishing the nature of the displacement w_3 on the wavefront, as will be shown in the next section. Note that, from (2.8.26), the sign of l is the negative of the normal curve curvature at the corresponding θ-point. Furthermore, since $p = R \cos \theta$ is the p-coordinate of a point on N, it follows that $(R \cos \theta)' > 0$ when p increases with increasing θ, while $(R \cos \theta)' < 0$ when p decreases with increasing θ. Thus the signs of l and n can be read off directly form a glance at the normal curve.

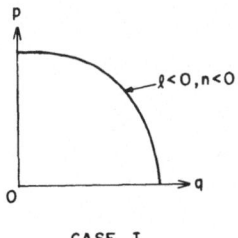

CASE I

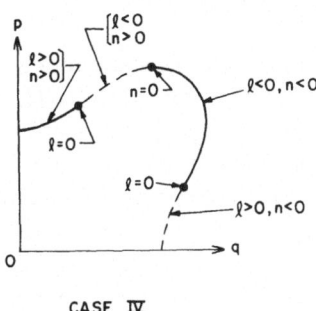

CASE IV

—— ELLIPTIC POINTS

• PARABOLIC POINTS

— — — HYPERBOLIC POINTS

Figure 62.   Classification of Points on the N_+ Sheet of the Slowness Surface. The p-axis is the axis of revolution.

According to Stoker [40] the surface N at a point can be classified as

Elliptic $ln > 0$, locally the surface lies entirely on one side of the tangent plane.
Parabolic $ln = 0$, locally the surface is a parabolic cylinder.
Hyperbolic $ln < 0$, locally the surface cuts the tangent plane along two intersecting lines.

These cases are illustrated in figure 62 for a Case I and Case IV section of the slowness surface. Only the N_+ sheet is drawn since the N_- sheet is composed entirely of elliptic points. Corresponding figures for Cases II, III and V can be supplied by the reader.

15. Remarks Concerning the Displacement Behaviour Near the Wavefront

By using the expression (3.14.25) for the slowness surface area element, together with the θ, ϕ parametric representation of N, the w_3 displacement in (3.13.52) may be

123

rendered more explicit

$$w_3 = w_{3s} + \frac{1}{4\pi^2} \int_0^{2\pi} d\phi \int_0^\pi d\theta\, F_-(\theta, \phi)\delta(\tau - \mathbf{x} \cdot \mathbf{q}_-)$$

$$- \frac{1}{4\pi^2} \int_0^{2\pi} d\phi \int_0^\pi d\theta\, F_+(\phi, \theta)\delta(\tau - \mathbf{x} \cdot \mathbf{q}_+), \tag{3.15.1}$$

where

$$F_\pm(\theta, \phi) = \frac{K(\mathbf{q}_\pm, 1)\sqrt{R_\pm^2 + R_\pm'^2}\, R_\pm \sin\theta}{|\nabla Q(\mathbf{q}_\pm, 1)|}. \tag{13.15.2}$$

A variety of things can be done with the above formula (and formulas of this type). Because of the transversely isotropic nature of the solid, $F_\pm$ is actually independent of the integration variable ϕ. Thus ϕ only appears explicitly, through (3.14.1), in the argument of the delta-function. By a suitable transformation the ϕ-integration can be performed, thereby showing that the solution is a function of the two space variables z and $\sqrt{x^2 + y^2}$. The remaining θ-integration is still formidable. However when attention is confined to the symmetry axis ($x = y = 0$), the entire integral is free of ϕ so that the ϕ-integration is trivial. This situation, approached differently, was investigated in section 3.11.

Several authors have made use of Herglotz–Petrowski formulas. Applications to crystal optics may be found in John [7], while Courant and Hilbert [18] used such formulas to locate the domain of dependence and lacunas for crystal optics. Both Duff [21, 24] and Burridge [53] have obtained asymptotic results valid near the wavefront for anisotropic elastic waves. Burridge [53], upon whose work the analysis of this section is based, was primarily concerned with the nature of the singular lid (caused by multiple points on the slowness surface) on the wavefront of cubic crystals. Barry and Musgrave [54] have made a similar investigation for elastic media with tetragonal symmetry. Related work, for harmonic waves, can be found in Buchwald [51] whose work is based on Lighthill [15] as well as Kline and Kay [19] who treat the asymptotic evaluation of certain electromagnetic diffraction integrals. Willis' [39] analysis of wavefront arrivals as well as the asymptotic evaluation of a Radon transform in Gel'fand, Graev and Vilenkin [55] are also relevant.

Consider now the behaviour of the integral

$$J = \frac{1}{4\pi^2} \int_0^{2\pi} d\phi \int_0^\pi d\theta\, F_+(\theta, \phi)\delta(\tau - \mathbf{x} \cdot \mathbf{q}_+) \tag{3.15.3}$$

for $\mathbf{x}$ and τ in the neighborhood of a point on the wavefront sheet W_+. Analysis of the second slowness surface integral in (3.15.1) can be similarly (and in fact more easily) handled. Now the only q-space points which contribute to the surface integration are those formed from the intersection of the slowness surface N and the plane $\tau = \mathbf{x} \cdot \mathbf{q}$. This locus of intersection points will in general consist of zero, one or two plane curves. According to Duff [24], 'As τ increases, the plane $\tau = \mathbf{x} \cdot \mathbf{q}$ moves away from the origin and an instant τ' at which it is tangent to a sheet $N_\pm$ of N is the moment the corresponding wave sheet (by the construction of section 2.18) $W_\pm$ of W reaches the point $\mathbf{x}$. As $\tau \to \tau'$ the intersection of the variable plane with N_+ or N_- (as the case may be), which is a 'vanishing cycle' in the sense of Petrowski [32], shrinks to a point and disappears.' Thus the singularity or jump in the solution on W involves only the q values attached to that point. Hence the singularities of w_3 on the sheet W_+ are contributed by a small patch of N_+ where the surface normal vector $\hat{n}_+$ is parallel to x [53]. In this regard, an exceptional case is that in which the tangent plane $\tau = \mathbf{x} \cdot \mathbf{q}$ touches N_+ along a closed curve (rotational version of the two-dimensional bitangent case) which corresponds to a conical point on W_+. This situation, and that for which N contains multiple points will not be treated here (see Burridge [53, 56]).

Suppose that the vanishing cycle on N_+ corresponding to the space vector x has θ, ϕ coordinates θ_0 and ϕ_0. The integrand $F_+(\theta, \phi)$ is then expanded about this point with only the zeroth approximation retained. Furthermore, writing q for $\mathbf{q}_+$,

$$\tau - \mathbf{x} \cdot \mathbf{q}(\theta, \phi) \approx \tau - \mathbf{x} \cdot \mathbf{q}(\theta_0, \phi_0) - \mathbf{x} \cdot \mathbf{q}_\theta(\theta_0, \phi_0)(\theta - \theta_0)$$
$$- \mathbf{x} \cdot \mathbf{q}_\phi(\theta_0, \phi_0)(\phi - \phi_0) - \tfrac{1}{2}\{\mathbf{x} \cdot \mathbf{q}_{\theta\theta}(\theta_0, \phi_0)(\theta - \theta_0)^2$$
$$+ 2\mathbf{x} \cdot \mathbf{q}_{\theta\phi}(\theta_0, \phi_0)(\theta - \theta_0)(\phi - \phi_0)$$
$$+ \mathbf{x} \cdot \mathbf{q}_{\phi\phi}(\theta_0, \phi_0)(\phi - \phi_0)^2\}. \tag{3.15.4}$$

But since

$$\mathbf{x} = |\mathbf{x}|\hat{n}_+, \tag{3.15.5}$$

equations (3.14.12) and (3.14.26) simplify (3.15.4) to

$$\tau - \mathbf{x} \cdot \mathbf{q}(\theta, \phi) \approx \tau - \mathbf{x} \cdot \mathbf{q}_+(\theta_0, \phi_0) - \frac{|\mathbf{x}|}{2}\{l(\theta - \theta_0)^2 + n(\phi - \phi_0)^2\}. \tag{3.15.6}$$

Thus

$$\delta(\tau - \mathbf{x} \cdot \mathbf{q}_+) \approx \frac{2}{|\mathbf{x}|} [c - l(\theta - \theta_0)^2 - n(\phi - \phi_0)^2]. \tag{3.15.7}$$

Hence

$$J \sim F_0 \int_0^{2\pi} d\phi \int_0^\pi d\theta \, \frac{2}{|\mathbf{x}|} \delta[c - l(\theta - \theta_0)^2 - n(\phi - \phi_0)^2], \tag{3.15.8}$$

where

$$F_0 \equiv \frac{1}{4\pi^2} F_+(\theta_0, \phi_0) \quad \text{and} \quad c = \frac{2}{|\mathbf{x}|}(\tau - \mathbf{x} \cdot \mathbf{q}_+(\theta_0, \phi_0)) \tag{3.15.9}$$

So, in the spirit of asymptotics,

$$J \sim \frac{2F_0}{|\mathbf{x}|} \int_{-\infty}^{+\infty}\int_{-\infty}^{+\infty} \delta(c - lu^2 - nv^2)du\,dv \tag{3.15.10}$$

where

$$|c| \ll 1. \tag{3.15.11}$$

The integral (3.15.10) will now be evaluated for three situations on the slowness surface,

(i) *Elliptic $l > 0$, $n > 0$*

Set

$$\sqrt{l}\,u = r\cos\theta, \qquad \sqrt{n}\,v = r\sin\theta, \tag{3.15.12}$$

then

$$J \sim \frac{2F_0}{|x|\sqrt{ln}} \int_0^\infty r\,dr \int_0^{2\pi} d\theta\,\delta(c - r^2) \tag{3.15.13}$$

or, with $r^2 = \zeta$

$$J \sim \frac{2\pi F_0}{|x|\sqrt{ln}} \int_0^\infty \delta(c - \zeta)\,d\zeta = \frac{2\pi F_0}{|x|\sqrt{ln}}\,H(c) \tag{3.15.14}$$

This corresponds to a jump in the displacement as the front W_+ is crossed.

(ii) *Elliptic $l < 0$, $n < 0$*

Set

$$\sqrt{-l}\,u = r\cos\theta, \qquad \sqrt{-n}\,v = r\sin\theta, \tag{3.15.15}$$

then

$$J \sim \frac{2F_0}{|x|\sqrt{ln}} \int_0^\infty r\,dr \int_0^{2\pi} d\theta\,\delta(c + r^2) = \frac{2\pi F_0}{|x|\sqrt{ln}}\,H(-c), \tag{3.15.16}$$

which again corresponds to a jump across W_+.

(iii) *Hyperbolic $l > 0$, $n < 0$*

Set

$$\sqrt{l}\,u = \bar{u}, \qquad \sqrt{-n}\,v = \bar{v} \tag{3.15.17}$$

then

$$J \sim \frac{2F_0}{|x|\sqrt{-ln}} \iint \delta(c - \bar{u}^2 + \bar{v}^2)\,d\bar{u}d\bar{v}. \tag{3.15.18}$$

But

$$\delta(c + \bar{v}^2 - \bar{u}^2) = \delta(\bar{u}^2 - c - \bar{v}^2)$$

$$= \frac{1}{2\sqrt{c + \bar{v}^2}}\,[\delta(\bar{u} - \sqrt{c + \bar{v}^2}) + \delta(\bar{u} + \sqrt{c + \bar{v}^2})]. \tag{3.15.19}$$

Suppose $c > 0$, then asymptotically, with a some convenient positive number,

$$J \sim \frac{2F_0}{|x|\sqrt{-ln}} \int_{-a}^{+a} d\bar{v} \int_{-\infty}^{+\infty} \frac{d\bar{u}}{2\sqrt{c + \bar{v}^2}}\,[\delta(\bar{u} - \sqrt{c + \bar{v}^2}) + \delta(\bar{u} + \sqrt{c + \bar{v}^2})]$$

$$\tag{3.15.20}$$

or

$$J \sim \frac{4F_0}{|\mathbf{x}|\sqrt{-ln}} \int_0^a \frac{d\bar{v}}{\sqrt{c + \bar{v}^2}} = \frac{4F_0}{|\mathbf{x}|\sqrt{-ln}} \sinh^{-1} a/\sqrt{c}. \tag{3.15.21}$$

But

$$\sinh^{-1} a/\sqrt{c} = \ln\left[\frac{a}{\sqrt{c}} + \sqrt{\frac{a^2}{c} + 1}\right] \sim \ln\frac{2a}{\sqrt{c}} \sim -\tfrac{1}{2}\ln c \tag{3.15.22}$$

as $c \to 0+$. Consequently

$$J \sim -\frac{2F_0}{|\mathbf{x}|\sqrt{-ln}} \ln c. \tag{3.15.23}$$

On the other hand if $c < 0$, call $B = -c$, interchanging the roles of $\bar{u}$ and $\bar{v}$ etc., give

$$J \sim -\frac{2F_0}{|\mathbf{x}|\sqrt{-ln}} \ln B. \tag{3.15.24}$$

Hence in general

$$J \sim -\frac{2F_0}{|\mathbf{x}|\sqrt{-ln}} \ln |c|. \tag{3.15.25}$$

The result (3.15.25) for the displacement behaviour near W_+ corresponding to hyperbolic points on N_+ was derived for $l > 0$ and $n < 0$. Obviously it remains true for $l < 0$ and $n > 0$. This logarithmic singularity, for the case of zinc was previously noticed by Cameron and Eason [44] in their investigation of wave motion in the transverse plane $z = 0$.

There is one minor difficulty when the above analysis is applied to the position vector $\mathbf{x} = (0, 0, z)$ since for this case $\theta_0 = 0$ and in view of (3.15.2)

$$F_\pm(0, \phi) = 0 \tag{3.15.26}$$

thus leading to incorrect asymptotic results. The trouble here, as mentioned in section 3.14, is caused by the particular coordinate system used. After replacing $\sin \theta$ by θ (as an approximation for small θ) everything will go through correctly. Since the axial displacement has played such a prominent role in section 3.11, a fresh derivation for the jump in w_3 on N_- at $z = \sqrt{\alpha}\tau$ will now be given. Here

$$q_-(0, \phi) = (0, 0, R_-(0)) = (0, 0, 1/\sqrt{\alpha}). \tag{3.15.27}$$

Call

$$J = \frac{1}{4\pi^2} \int_0^{2\pi} d\phi \int_0^\pi F_-(\theta, \phi) \delta(\tau - \mathbf{x} \cdot \mathbf{q}_-), \tag{3.15.28}$$

so

$$J \sim \frac{1}{2\pi} \int_0^\pi \theta d\theta \, U(\mathbf{q}_-) \delta(\tau - zR_- \cos \theta) \tag{3.15.29}$$

where

$$U(q_-) = \left. \frac{K(q_-, 1)\sqrt{R_-^2 + R_-'^2}\,R_-}{|\nabla Q(q_-, 1)|} \right|_{q_- = (0, 0, 1/\sqrt{\alpha})} \tag{3.15.30}$$

A little calculation will show that, for the above value of q_-

$$K = -(\alpha - 1)/\alpha, \qquad |\nabla Q| = |Q_p| = 2(\alpha - 1)\alpha^{-1/2}$$

$$R_-(0) = \alpha^{-1/2} \quad \text{and} \quad R_-'(0) = 0$$

so

$$U = -\frac{1}{2\alpha^{3/2}}$$

and

$$J \sim -\frac{1}{4\pi\alpha^{3/2}} \int_0^\pi \delta\left(\tau - zR_-(\theta)\cos\theta\right)\theta d\theta \tag{3.15.31}$$

Using

$$R_-''(0) = \frac{\alpha^2 - \beta\alpha - 2\alpha + \gamma}{\alpha^{3/2}(\alpha - 1)} \tag{3.15.32}$$

where $R_-(\theta)$ was defined in (2.16.10), then to second order in θ

$$\tau - zR_-(\theta)\cos\theta \approx (\tau - z/\sqrt{\alpha}) + \frac{z(\alpha + \beta\alpha - \gamma)}{2(\alpha - 1)\alpha^{3/2}}\theta^2 \equiv C + D\theta^2. \tag{3.15.33}$$

Then

$$J \sim -\frac{1}{4\pi\alpha^{3/2}} \int_0^\infty \delta(C + D\theta^2)\theta d\theta \tag{3.15.34}$$

or, calling $\theta^2 = s$

$$J \sim -\frac{1}{8\pi\alpha^{3/2}|D|} \int_0^\infty \delta\left(\frac{C}{D} + s\right) ds, \tag{3.15.35}$$

$$J \sim -\frac{H(-C)}{8\pi\alpha^{3/2}D}$$

since $D > 0$ for $\alpha > 1$, $z > 0$.

Consequently

$$J \sim -\frac{1}{4\pi z}\frac{\alpha - 1}{(\beta + 1)\alpha - \gamma}H(z/\tau - \sqrt{\alpha}). \tag{3.15.36}$$

Now the static part of w_3 (namely w_{3s}) and the second slowness surface integral in (3.15.1) will remain continuous as the front $\tau = z/\sqrt{\alpha}$ crosses the point z on the (positive) symmetry axis. In view of this, and the fact that $w_3(0, 0, z, \tau) \equiv 0$ prior to the arrival of this leading sheet of the wavefront, it is seen that the asymptotic result (3.15.36) is in complete agreement with the magnitude (and sign) of the jump predicted by the exact analysis of section 3.11 and in particular equation (3.11.40).

Incidentally the solution will remain continuous at the front $z = \tau$ since $K(0, 0, 1, 1) = 0$.

4

Surface motion of a two-dimensional half-space (Lamb's Problem)

In a famous paper published in 1904, Lamb [57] completely solved the problem of the surface motion of a two-dimensional isotropic half-space subjected to a point impulsive normal load applied at the surface. Lamb's paper contains a rich variety of other results, but it is the transversely isotropic version of the half-space problem that forms the subject matter of the present chapter. Following the penetrating treatment of this problem by Kraut [58], the axis of material symmetry (z-axis) will be assumed perpendicular to the surface of the half-space. Unlike the isotropic problem solved by Lamb, the orientation of the crystallographic axes relative to the half-space surface does make a considerable difference in the complexity of the analysis for the corresponding transversely isotropic solid. Burridge [59] has analyzed Lamb's problem for a general anisotropic solid in which the surface has no particularly symmetrical orientation with respect to the crystallographic axes of the medium. Burridge illustrates in some detail his results (which are not explicit) for the cubic crystal copper. On the other hand the problem formulated by Kraut (and illustrated for the crystal Beryl) and treated later by Budaev [60], does allow an explicit closed form solution, although neither author takes the opportunity to do so.

As in chapters 2 and 3, the focus of attention in chapter 4 will be on those aspects of the problem, in this case Lamb's problem, which admit an explicit closed form solution. Here these conditions are satisfied for the surface displacements, which are herein treated in detail for the 13 crystals listed in table 7, and for the displacements along the epicentral axis beneath the applied surface load. The latter displacements will not be treated in this chapter, but will instead be studied in the more general three-dimensional context in chapter 5. Kraut [58] has also noticed the special nature of the epicentral axis displacements and gives graphical results for Beryl for the two-dimensional half-space problem.

1. Formulation of the Problem

The half-space $z > 0$, in which the symmetry axis is normal to the surface $z = 0$, is impulsively loaded by the surface traction **T**. The problem is assumed to be one of plane strain with the displacement vector **u** represented by

$$\mathbf{u} = (0, v, w) \quad \text{where} \quad v = v(y, z, t) \quad \text{and} \quad w = w(y, z, t). \qquad (4.1.1)$$

The traction vector, which is applied normal to the surface and at the origin, is given

130

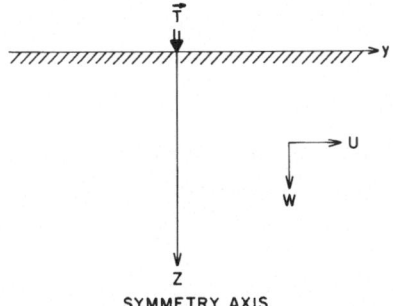

Figure 63.   Surface Point Loading of a Two-Dimensional Half-Space

by

$$\mathbf{T} = P_0(0, 0, \delta(y)\delta(t)), \tag{4.1.2}$$

where P_0 measures the magnitude of the applied load. In Cartesian tensor notation, the traction vector is related to the stress tensor by

$$T_i = \sigma_{ij}n_j \quad \text{with} \quad n_j = (0, 0, -1). \tag{4.1.3}$$

Equation (4.1.3) is identically satisfied when $i = 1$, since from (3.7.1) $\sigma_{13} \equiv 0$ under the condition of plane strain. Taking $i = 2$ and 3 in (4.1.3) gives the two surface boundary conditions

$$0 = -c_{44}\left[\frac{\partial v}{\partial z} + \frac{\partial w}{\partial y}\right] \quad \text{and} \quad P_0\delta(y)\delta(t) = -\left[c_{13}\frac{\partial v}{\partial y} + c_{33}\frac{\partial w}{\partial z}\right].$$

$$\tag{4.1.4}$$

Introducing dimensionless material parameters, from (1.5.22) and (1.6.4) and a stretched time variable τ from (1.5.23) allows the boundary conditions to be written as

$$\frac{\partial v}{\partial z} + \frac{\partial w}{\partial y} = 0 \quad \text{on} \quad z = 0, \tag{4.1.5}$$

$$(\kappa - 1)\frac{\partial v}{\partial y} + \alpha\frac{\partial w}{\partial z} = -\frac{P_0}{\sqrt{c_{44}\rho}}\,\delta(y)\delta(\tau) \quad \text{on} \quad z = 0 \tag{4.1.6}$$

and

$$v \to 0, w \to 0 \quad \text{as} \quad \sqrt{y^2 + z^2} \to \infty. \tag{4.1.7}$$

The equations of motion (1.5.10), with zero body force and $u \equiv 0$, $\partial/\partial x \equiv 0$ reduce to

$$\left[\beta\frac{\partial^2}{\partial y^2} + \frac{\partial^2}{\partial z^2} - \frac{\partial^2}{\partial \tau^2}\right][v] + \left[\kappa\frac{\partial^2}{\partial y\partial z}\right][w] = 0, \tag{4.1.8}$$

and

$$\left[\kappa\frac{\partial^2}{\partial y\partial z}\right][v] + \left[\frac{\partial^2}{\partial y^2} + \alpha\frac{\partial^2}{\partial z^2} - \frac{\partial^2}{\partial \tau^2}\right][w] = 0. \tag{4.1.9}$$

The central problem of this chapter is concerned with solving equations (4.1.8) and (4.1.9) subject to zero initial conditions and the boundary conditions (4.1.5)–(4.1.7), in the domain of the half-space $-\infty < y < \infty, z \geqslant 0$.

2. Integral Transform Representation of the Solution When the Fourier Inversion Path is Free of Branch Points

Define the Fourier-Laplace transform of v by

$$\bar{v}^*(\eta, z, s) = \int_{-\infty}^{+\infty} \int_0^\infty v(y, z, \tau) e^{-i\eta y - s\tau} d\tau dy,$$ (4.2.1)

with inverse

$$v(y, z, \tau) = \frac{1}{(2\pi)^2 i} \int_{-\infty}^{+\infty} \int_c \bar{v}^*(\eta, z, s) e^{i\eta y + s\tau} ds d\eta,$$ (4.2.2)

with similar expressions for w and $\bar{w}^*$. Taking the transform of the equations of motion (4.1.8) and (4.1.9) gives

$$\left[-\beta\eta^2 + \frac{\partial^2}{\partial z^2} - s^2\right][\bar{v}^*] + i\eta\kappa \frac{\partial}{\partial z}[\bar{w}^*] = 0,$$ (4.2.3)

$$i\eta\kappa \frac{\partial}{\partial z}[\bar{v}^*] + \left[-\eta^2 + \alpha\frac{\partial^2}{\partial z^2} - s^2\right][\bar{w}^*] = 0,$$ (4.2.4)

while the boundary conditions (4.1.5) and (4.1.6) transform to

$$\frac{\partial \bar{v}^*}{\partial z} + i\eta\bar{w}^* = 0 \quad \text{on} \quad z = 0,$$ (4.2.5)

$$i\eta(\kappa - 1)\bar{v}^* + \alpha\frac{\partial \bar{w}^*}{\partial z} = -\frac{P_0}{\sqrt{c_{44}\rho}} \quad \text{on} \quad z = 0.$$ (4.2.6)

Solutions to the homogeneous system of first order ordinary differential equations are now sought in the form

$$\bar{v}^* = A e^{kz} \quad \text{and} \quad \bar{w}^* = B e^{kz}$$ (4.2.7)

where, of course, A, B and k are independent of the variable z. Insertion of (4.2.7) into the system (4.2.3) and (4.2.4) gives

$$(-\beta\eta^2 + k^2 - s^2)A + i\eta\kappa kB = 0,$$ (4.2.8)

$$i\eta\kappa kA + (-\eta^2 + \alpha k^2 - s^2)B = 0,$$ (4.2.9)

Necessary and sufficient conditions for the existence of a solution to these equations is that the determinant of the coefficients of A and B must vanish, i.e.,

$$\alpha k^4 - k^2[(\alpha + 1)s^2 + \gamma\eta^2] + [\beta\eta^4 + (\beta + 1)s^2\eta^2 + s^4] = 0.$$ (4.2.10)

In the spirit of Lamb [57], make the substitutions

$$k = s\zeta \quad \text{and} \quad \eta = s\omega \tag{4.2.11}$$

where the Laplace transform variable s may be taken to be real and positive due to a real inversion formula of Widder [61]. Then the above quartic (biquadratic) for k becomes

$$\alpha\zeta^4 - \zeta^2[(\alpha+1)+\gamma\omega^2] + [\beta\omega^4 + (\beta+1)\omega^2 + 1] = 0, \tag{4.2.12}$$

with ζ^2-roots

$$\zeta^2 = \{[\gamma\omega^2 + (\alpha+1)] \pm [(\gamma\omega^2 + \alpha + 1)^2 - 4\alpha(\beta\omega^4 + (\beta+1)\omega^2 + 1)]^{1/2}\}/2\alpha. \tag{4.2.13}$$

Call

$$\zeta_1(\omega) = \{\gamma\omega^2 + (\alpha+1) + \sqrt{\phi(\omega)}\}^{1/2}/\sqrt{2\alpha},$$

$$\zeta_2(\omega) = -\zeta_1(\omega),$$

$$\zeta_3(\omega) = \{\gamma\omega^2 + (\alpha+1) - \sqrt{\phi(\omega)}\}^{1/2}/\sqrt{2\alpha}$$

and

$$\zeta_4(\omega) = -\zeta_3(\omega) \tag{4.2.14}$$

where

$$\phi(\omega) = [\gamma\omega^2 + (\alpha+1)]^2 - 4\alpha[\beta\omega^4 + (\beta+1)\omega^2 + 1]. \tag{4.2.15}$$

If the ζ-quartic (4.2.12) is to have real roots, the following conditions must be imposed

(i)   if $\alpha > \beta > 1$ then $2\sqrt{\alpha\beta} \leqslant \gamma \leqslant (1+\alpha\beta)$,

(ii)   if $\beta > \alpha > 1$ then $(\alpha+\beta) \leqslant \gamma \leqslant (1+\alpha\beta)$,   (4.2.16)

(iii)   if $\alpha = \beta > 1$ then $2\alpha \leqslant \gamma \leqslant (1+\alpha^2)$.

Note that only materials for which $\alpha > 1$ and $\beta > 1$ (as is true for the crystals of Table 7) have been considered in the constraints (4.2.16). However not all the crystals listed in Tables 7 and 11 satisfy (4.2.16), in fact only the nine crystals denoted by a (1) in Table 11, meet these constraints. The slowness curves for these nine crystals are locally convex (class I or V from Table 7) where the slowness axis, which is perpendicular to the symmetry axis, pierces the curve. This implies that no triangular portions of the wave front will intersect the material surface $z = 0$. The class of materials obeying (4.2.16) includes isotropic materials so that it might be anticipated that the surface displacements (as opposed to the interior displacements for class V materials) will share some features with those of an isotropic solid. Conditions (4.2.16) will be relaxed below in sections 4 and 5 of this chapter, so as to include the remaining four crystals listed in Table 11.

Before exploiting the implications of conditions (4.2.16) the (possible) branch point locations in the complex ω-plane will be examined. Such branch points for $\zeta(\omega)$

may arise from (4.2.13) in two distinct ways

(1) ω-values for which $\phi(\omega) = 0$,   (4.2.17)

and

(2) ω-values for which $\gamma\omega^2 + (\alpha + 1) \pm \sqrt{\phi(\omega)} = 0$ (4.2.18)

Now (1) implies

$$\omega^2 = \{2\alpha(\beta + 1) - \gamma(\alpha + 1)$$
$$\pm i\,[4\alpha\{\gamma - (\alpha + \beta)\}\,\{(1 + \alpha\beta) - \gamma\}]^{1/2}\}/(\gamma^2 - 4\alpha\beta) \qquad (4.2.19)$$

such ω-points being complex, while (2) implies (note that a squaring may introduce extraneous roots here)

$$\omega = \pm i/\sqrt{\beta} \quad \text{and} \quad \omega = \pm i. \qquad (4.2.20)$$

In either event it is clear that the imposition of conditions (4.2.16) implies that the real ω-axis is free of branch points.

Because of the above $\zeta(\omega)$ branch point locations, it is now clear that (and this is the basis for assuming conditions (4.2.16))

$$\zeta_1(\omega) > 0, \qquad \zeta_3(\omega) > 0; \qquad \zeta_2(\omega) < 0, \qquad \zeta_4(\omega) < 0 \qquad (4.2.21)$$

for $-\infty < \omega < +\infty$. Then, because of the boundary condition (4.1.7) at infinity, only $\zeta_2(\omega)(= -\zeta_1)$ and $\zeta_4(\omega)(= -\zeta_3)$ are suitable exponents for $\bar{v}^*$ and $\bar{w}^*$. Using now the fact that $\bar{v}^*$ and $\bar{w}^*$ will (separately) be a linear combination of $e^{-s\zeta_1 z}$ $e^{-s\zeta_3 z}$ and that the coefficients of each exponential term in $\bar{v}^*$ and $\bar{w}^*$ are not independent, but related through equations (4.2.8) and (4.2.9), gives

$$\bar{v}^* = \frac{\alpha\zeta_1^2 - \omega^2 - 1}{i\kappa\omega\zeta_1} A_1 e^{-s\zeta_1 z} + \frac{\alpha\zeta_3^2 - \omega^2 - 1}{i\kappa\omega\zeta_3} C e^{-s\zeta_3 z} \qquad (4.2.22)$$

and

$$\bar{w}^* = A_1 e^{-s\zeta_1 z} + C e^{-s\zeta_3 z} \qquad (4.2.23)$$

When these expressions are substituted into the boundary conditions (4.2.5) and (4.2.6) a pair of simultaneous equations for A_1 and C results whose solution is readily found to be

$$A_1 = \frac{P_0}{s\sqrt{c_{44}\rho}} \frac{\zeta_1\zeta_3}{\zeta_1 - \zeta_3} \frac{\alpha\zeta_3^2 + (\kappa - 1)\omega^2 - 1}{D}, \qquad (4.2.24)$$

and

$$C = -\frac{P_0}{s\sqrt{c_{44}\rho}} \frac{\zeta_1\zeta_3}{\zeta_1 - \zeta_3} \frac{\alpha\zeta_1^2 + (\kappa - 1)\omega^2 - 1}{D}, \qquad (4.2.25)$$

where the Rayleigh function D is given by

$$D(\omega) = [2(1 - \kappa)\omega^2(\omega^2 + 1) - (\gamma\omega^2 + \alpha)(\omega^2 + 1) - \alpha\zeta_1\zeta_3]. \qquad (4.2.26)$$

Equations (4.2.22)–(4.2.26) constitute the (transformed) solution to the problem posed in this section. The inversion of these expressions at points other than on the epicentral-axis ($y = 0$) or the surface ($z = 0$) is very complicated (see Kraut [58] for an indication of the difficulties). For this reason only the surface displacements will be treated herein. On the surface

$$\bar{v}^*(\eta, 0, s) = -\frac{P_0}{s\sqrt{c_{44}\rho}} \frac{i\omega\{\alpha\varsigma_1\varsigma_3 + (1 - \kappa)(\omega^2 + 1)\}}{D}, \qquad (4.2.27)$$

and

$$\bar{w}^*(\eta, 0, s) = -\frac{P_0\alpha}{s\sqrt{c_{44}\rho}} \frac{\varsigma_1\varsigma_3(\varsigma_1 + \varsigma_3)}{D}. \qquad (4.2.28)$$

Note from the above that both $\bar{v}^*(\eta, 0, s)$ and $\bar{w}^*(\eta, 0, s)$ are even functions of $\sqrt{\phi(\omega)}$ so that the branch points arising from 1) in equation (4.2.17) may be neglected. These points, from (4.2.19), will play a future role in section 4.5.

Before closing this section, asymptotic expressions as $|\omega| \to \infty$ will be listed for $\bar{v}^*$ and $\bar{w}^*$

$$\bar{v}^*(\eta, 0, s) \sim -\frac{P_0}{s\sqrt{c_{44}\rho}} \frac{i[\sqrt{\alpha\beta} + 1 - \kappa]}{2(1 - \kappa) - \gamma} \cdot \frac{1}{\omega}, \qquad (4.2.29)$$

$$\bar{w}^*(\eta, 0, s) \sim$$

$$-\frac{P_0\alpha}{s\sqrt{c_{44}\rho}} \sqrt{\frac{\beta}{\alpha}} \frac{[\gamma + (\gamma^2 - 4\alpha\beta)^{1/2}]^{1/2}/\sqrt{2\alpha} + [\gamma - (\gamma^2 - 4\alpha\beta)^{1/2}]^{1/2}/\sqrt{2\alpha}}{2(1 - \kappa) - \gamma} \cdot \frac{1}{\omega}$$

$$(4.2.30)$$

These expressions will be useful in later sections when the Fourier integration path is deformed into an alternate integration path.

Finally when the inverse Fourier transform part of formula (4.2.2) is applied to (4.2.27) and (4.2.28), noting that $\bar{v}^*(\eta, 0, s)$ is odd in ω while $\bar{w}^*(\eta, 0, s)$ is even, together with (4.2.11), gives

$$\bar{w}(y, 0, s) = -\frac{P_0\alpha}{\sqrt{c_{44}\rho}} \operatorname{Re} \frac{1}{\pi} \int_0^\infty \frac{\varsigma_1\varsigma_3(\varsigma_1 + \varsigma_3)}{D} e^{is\omega|y|} d\omega \qquad (4.2.31)$$

and

$$\bar{v}(y, 0, s) = \frac{P_0}{\sqrt{c_{44}\rho}} \operatorname{sgn} y \operatorname{Im} \frac{1}{\pi} \int_0^\infty \frac{\omega[\alpha\varsigma_1\varsigma_3 + (1 - \kappa)(\omega^2 + 1)]}{D} e^{is\omega|y|} d\omega \qquad (4.2.32)$$

where Re indicates the real part, Im indicates the imaginary part and sgn y stands for the sign of y.

3. Transform Inversion for Materials Satisfying Condition (1) of Table 11

The treatment in this section of the integrals (4.2.31) and (4.2.32) will not be exhaustive insofar as the conditions (4.2.16) are concerned, but it will be general enough to include all nine crystals designated by (1) in Table 11. For these nine crystals either (i) or (ii) of conditions (4.2.16) applies, but always $\gamma > (\alpha + \beta)$.

It was mentioned earlier in section (4.2) (above equation (4.2.20)) that $\omega = \pm i$ and $\omega = \pm i/\sqrt{\beta}$ may not all be branch points of $\zeta_1(\omega)$ and $\zeta_3(\omega)$. These points will now be individually investigated. From equation (4.2.14)

$$\zeta_1(i) = \{-\gamma + (\alpha + 1) + \sqrt{(-\gamma + \alpha + 1)^2}\}^{1/2}/\sqrt{2\alpha},$$

but $\gamma > (\alpha + 1)$ for (1) materials so

$$\sqrt{(-\gamma + \alpha + 1)^2} = \gamma - (\alpha + 1) \quad \text{hence} \quad \zeta_1(i) = 0 \quad \text{and}$$

$$\omega = i \quad (\text{and } \omega = -i) \text{ is a branch point of } \zeta_1(\omega). \tag{4.3.1}$$

Similarly

$$\omega = i \quad (\text{and } \omega = -i) \text{ is not a branch point of } \zeta_3(\omega). \tag{4.3.2}$$

Also from equation (4.2.14)

$$\zeta_1(i/\sqrt{\beta}) = \{-\gamma/\beta + (\alpha + 1) + \sqrt{(-\gamma/\beta + \alpha + 1)^2}\}^{1/2}/\sqrt{2\alpha},$$

but $-\gamma/\beta + (\alpha + 1) = (\beta + \alpha\beta - \gamma)/\beta > 0$ since $\gamma \leqslant 1 + \alpha\beta < \beta + \alpha\beta$. Therefore $\zeta_1(i/\sqrt{\beta}) \neq 0$ so,

$$\omega = i/\sqrt{\beta} \quad (\text{and } \omega = -i/\sqrt{\beta}) \text{ is not a branch point of } \zeta_1(\omega). \tag{4.3.3}$$

Similarly

$$\omega = i/\sqrt{\beta} \quad (\text{and } \omega = -i/\sqrt{\beta}) \text{ is a branch point of } \zeta_3(\omega). \tag{4.3.4}$$

It is interesting to note that the existence of (possible) branch points on the imaginary axis of the complex ω-plane, as indicated in equations (4.3.1)–(4.3.4), is in complete agreement with the isotropic case where $\alpha = \beta = \gamma/2$ and

$$\zeta_1(\omega) = \sqrt{\omega^2 + 1}, \qquad \zeta_3(\omega) = \sqrt{\omega^2 + 1/\beta}. \tag{4.3.5}$$

Anticipating that the main interest in this section will be the behaviour of $\zeta_1(\omega)$ and $\zeta_3(\omega)$ on the positive imaginary ω-axis (which is the Cagniard-Pekeris inversion path for the surface displacements) it is seen that

$$\phi(i\omega_2) > 0 \tag{4.3.6}$$

where

$$\omega = \omega_1 + i\omega_2. \tag{4.3.7}$$

The inequality (4.3.6) follows from (4.2.19) where it was shown that $\phi(\omega) = 0$ has only complex roots of the form $\omega = a + ib$ with $a \neq 0$ (at least for the nine crystals listed as (1) in table 11) hence $\phi(i\omega_2) \neq 0$, and hence can not change sign on $\omega \doteq i\omega_2$. But as $|\omega| \to \infty$, $\phi(\omega) \sim (\gamma^2 - 4\alpha\beta)\omega^4 > 0$ for $\gamma > 2\sqrt{\alpha\beta}$, thus (4.3.6) is justified. Also

$$\zeta_1^2(i\omega_2) > 0 \quad \text{for} \quad 0 \leqslant \omega_2 < 1 \quad \text{and} \quad \zeta_1^2(i\omega_2) < 0 \quad \text{for} \quad \omega_2 > 1.$$

$$\tag{4.3.8}$$

These inequalities are true since $\zeta_1^2(0) = 1$, $\zeta_1^2(i) = 0$ and $\zeta_1^2(i\omega_2) \sim$ $(-\gamma + \sqrt{\gamma^2 - 4\alpha\beta})\omega_2^2/2\alpha < 0$ as $\omega_2 \to \infty$. Similarly

$$\zeta_3^2(i\omega_2) > 0 \quad \text{for} \quad 0 \leqslant \omega_2 < 1/\sqrt{\beta} \quad \text{and} \quad \zeta_3^2(i\omega_2) < 0 \quad \text{for} \quad \omega_2 > 1/\sqrt{\beta} .$$

(4.3.9)

The behaviour of the product

$$\zeta_1\zeta_3 = \sqrt{\frac{\beta}{\alpha}} \sqrt{(\omega^2 + 1/\beta)(\omega^2 + 1)}$$

(4.3.10)

in the complex ω-plane plays an important part in the analysis of this chapter. Branch cuts for this expression will be taken on $\omega = \omega_1 \pm i/\sqrt{\beta}$ with $\omega_1 \leqslant 0$ and $\omega = \omega_1 \pm i$ with $\omega_1 \leqslant 0$ (see figure 64).

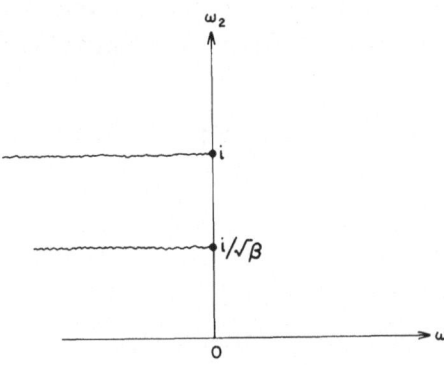

Figure 64. Branch Cuts for $\zeta_1\zeta_3$ in the Upper Half ω-Plane

These branch cuts, which make $\zeta_1\zeta_3$ single-valued in the cut ω-plane, are defined so that

$$\omega - i/\sqrt{\beta} = R_1 e^{i\theta_1}; \qquad R_1 = |\omega - i/\sqrt{\beta}|, \qquad \theta_1 = \arg(\omega - i/\sqrt{\beta}),$$
$$-\pi < \theta_1 < \pi,$$

$$\omega + i/\sqrt{\beta} = R_2 e^{i\theta_2}, \qquad R_2 = |\omega + i/\sqrt{\beta}|, \qquad \theta_2 = \arg(\omega + i/\sqrt{\beta}),$$
$$-\pi < \theta_2 < \pi,$$

then

$$\sqrt{\omega^2 + 1/\beta} = (R_1 R_2)^{1/2} e^{i(\theta_1 + \theta_2)/2},$$

(4.3.11)

and

$$\omega - i = R_3 e^{i\theta_3}, \qquad R_3 = |\omega - i|, \qquad \theta_3 = \arg(\omega - i),$$
$$-\pi < \theta_3 < \pi,$$

$$\omega + i = R_4 e^{i\theta_4}, \qquad R_4 = |\omega + i|, \qquad \theta_4 = \arg(\omega + i),$$
$$-\pi < \theta_4 < \pi,$$

then

$$\sqrt{\omega^2 + 1} = (R_3 R_4)^{1/2} e^{i(\theta_3 + \theta_4)/2}. \tag{4.3.12}$$

With the function $\zeta_1 \zeta_3$ so defined, it follows that on the positive ω_2-axis,

$$\zeta_1 \zeta_3 = \sqrt{\frac{\beta}{\alpha}} \sqrt{(1/\beta - \omega_2^2)(1 - \omega_2^2)} \quad \text{for} \quad 0 \leqslant \omega_2 \leqslant 1/\sqrt{\beta}, \tag{4.3.13}$$

$$\zeta_1 \zeta_3 = i\sqrt{\frac{\beta}{\alpha}} \sqrt{(\omega_2^2 - 1/\beta)(1 - \omega_2^2)} \quad \text{for} \quad 1/\sqrt{\beta} \leqslant \omega_2 \leqslant 1, \tag{4.3.14}$$

and

$$\zeta_1 \zeta_3 = -\sqrt{\frac{\beta}{\alpha}} \sqrt{(\omega_2^2 - 1/\beta)(\omega_2^2 - 1)} \quad \text{for} \quad \omega_2 \geqslant 1. \tag{4.3.15}$$

In view of the above expressions for $\zeta_1 \zeta_3$ together with the facts regarding the sign of $\zeta_1^2(i\omega_2)$ and $\zeta_3^2(i\omega_2)$ indicated in equations (4.3.8) and (4.3.9), it is consistent to take

$$\zeta_1(i\omega_2) = \begin{cases} i|\zeta_1(i\omega_2)| & \text{for} \quad \omega_2 \geqslant 1 \\ |\zeta_1(i\omega_2)| & \text{for} \quad 0 \leqslant \omega_2 \leqslant 1, \end{cases} \tag{4.3.16}$$

and

$$\zeta_3(i\omega_2) = \begin{cases} i|\zeta_3(i\omega_2)| & \text{for} \quad \omega_2 \geqslant 1/\sqrt{\beta} \\ |\zeta_3(i\omega_2)| & \text{for} \quad 0 \leqslant \omega_2 \leqslant 1/\sqrt{\beta}. \end{cases} \tag{4.3.17}$$

Or, more explicitly from (4.2.14)

$$\zeta_1(i\omega_2) = i\{\gamma\omega_2^2 - (\alpha + 1) - \sqrt{\phi(i\omega_2)}\}^{1/2}/\sqrt{2\alpha} \quad \text{for} \quad \omega_2 \geqslant 1, \tag{4.3.18}$$

$$\zeta_1(i\omega_2) = \{-\gamma\omega_2^2 + (\alpha + 1) + \sqrt{\phi(i\omega_2)}\}^{1/2}/\sqrt{2\alpha} \quad \text{for} \quad 0 \leqslant \omega_2 \leqslant 1, \tag{4.3.19}$$

$$\zeta_3(i\omega_2) = i\{\gamma\omega_2^2 - (\alpha + 1) + \sqrt{\phi(i\omega_2)}\}^{1/2}/\sqrt{2\alpha} \quad \text{for} \quad \omega_2 \geqslant 1/\sqrt{\beta}, \tag{4.3.20}$$

and

$$\zeta_3(i\omega_2) = \{-\gamma\omega_2^2 + (\alpha + 1) - \sqrt{\phi(i\omega_2)}\}^{1/2}/\sqrt{2\alpha} \quad \text{for} \quad 0 \leqslant \omega_2 \leqslant 1/\sqrt{\beta}. \tag{4.3.21}$$

Regarding the above expressions for $\zeta_1(i\omega_2)$ and $\zeta_3(i\omega_2)$ recall that $\phi(\omega)$ is defined in (4.2.15) and that $\phi(i\omega_2)$ is positive from (4.3.6).

The discussion of the ω-plane singularities for the surface displacement integrals (4.2.31) and (4.2.32) is now complete, save for the Rayleigh function $D(\omega)$ defined in (4.2.26). The branch points of this function have been treated above, so that $D(\omega)$ need only be examined for possible zeros which would correspond to pole singularities. By inspection of (4.2.26) and (4.3.10) it is evident that $D(\pm i) = 0$ but these zeros correspond to branch point rather than pole singularities of $1/D(\omega)$. After squaring and simplifying, the equation $D(\omega) = 0$ yields a cubic equation in ω^2

$$a^2\omega^6 + a(a + 2\alpha)\omega^4 + \alpha(2a + \alpha - \beta)\omega^2 + \alpha(\alpha - 1) = 0, \qquad (4.3.22)$$

where $a = 2\kappa - 2 + \gamma$.

The relevant (upper half ω-plane) root of (4.3.22) will fall on the imaginary ω-axis with $\omega_2 > 1$. (Stonely [62]). Because of the way in which the branch cuts for $\zeta_1\zeta_2$ were defined in (4.3.11) and (4.3.12) the remaining roots of (4.3.22) will lie on nonphysical sheets of the associated Riemann surface. There is analytic evidence for an isotropic solid that the remaining zeros of (4.3.22) '...that are on a forbidden sheet can yet make an identifiable contribution on a seismogram,' according to Aki and Richards [63]. In this connection reference should also be made to Phinney [64], Gilbert and Laster [65] and Chapman [66]. No attempt has yet been made to explore this interesting question for anisotropic media.

Denote the (Rayleigh pole) root of (4.3.22) by $i\omega_R$ so that

$$D(i\omega_R) = 0. \qquad (4.3.23)$$

Table 12 gives numerical values for $i/\sqrt{\beta}$ and $i\omega_R$ for the nine crystals of Table 11 which satisfy $(\alpha + \beta) < \gamma < (1 + \alpha\beta)$.

Table 12.   Surface Displacement Branch Point and Rayleigh Pole Location in the Upper Half ω-Plane for the (1) Crystals of Table 11.

Crystal	α	β	γ	i	$i/\sqrt{\beta}$	$i\omega_R$
Beryl	3.62	4.11	11.81	i	$i0.49326$	$i1.04649$
Cobalt	4.74	4.07	14.69	i	$i0.49568$	$i1.03976$
Hafnium	3.54	3.25	7.72	i	$i0.55470$	$i1.07193$
Ice	4.57	4.26	13.51	i	$i0.48450$	$i1.04333$
Magnesium	3.74	3.61	9.20	i	$i0.52632$	$i1.06098$
Rhenium	4.22	3.78	11.77	i	$i0.51434$	$i1.04888$
Thallium	7.27	5.62	16.93	i	$i0.42182$	$i1.04581$
Titanium	3.88	3.47	8.31	i	$i0.53683$	$i1.06871$
Yttrium	3.16	3.20	7.81	i	$i0.55902$	$i1.06954$

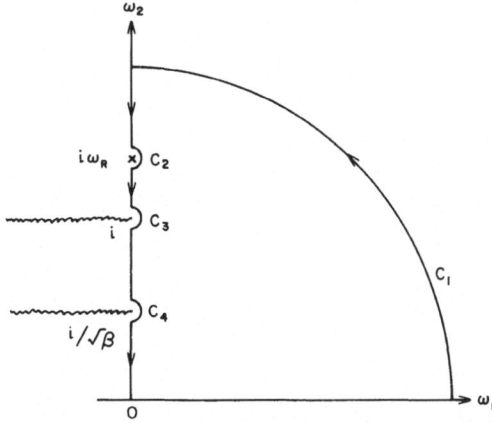

Figure 65.   Complex ω-Plane Showing the Contours Used in Evaluating the Surface Displacements for (1) Materials of Table 11.

139

The necessary ground work has now been completed for the integration contour deformation and subsequent inversion of integrals (4.2.31) and (4.2.32). Since the exponents for these integrals are real on the imaginary ω-axis, the Cagniard-Pekeris path is the positive ω_2-axis. Figure 65 indicates how the original ω-plane integration along the positive ω_1-axis is linked to the Cagniard-Pekeris path.

w inversion

Call

$$W(\omega) = \tilde{W}(\omega)e^{is\omega|y|}, \tag{4.3.24}$$

where

$$\tilde{W}(\omega) = \frac{\zeta_1(\omega)\zeta_3(\omega)\,[\zeta_1(\omega) + \zeta_3(\omega)]}{D(\omega)} \tag{4.3.25}$$

so that $W(\omega)$ is the integrand of (4.2.31) to within a multiplicative constant. Now integrate $W(\omega)$ around the closed contour shown in figure 65. By Cauchy's (1825) theorem, this integral will vanish since no singularities are enclosed.

$$\int_0^\infty W(\omega_1)d\omega_1 + \int_{c_1} + P\int_\infty^1 W(i\omega_2)id\omega_2 - \pi i \{\text{Residue } W(\omega) \text{ at } \omega = i\omega_R\}$$

$$+ \int_{c_3} + \int_1^{1/\sqrt{\beta}} W(i\omega_2)id\omega_2 + \int_{c_4} + \int_{1/\sqrt{\beta}}^0 W(i\omega_2)id\omega_2$$

$$= 0, \tag{4.3.26}$$

where the P in front of the third integral indicates that this integral is to be interpreted as a Cauchy principle value integral. From (4.2.30) and Jordan's lemma

$$\int_{c_1} \to 0 \quad \text{as} \quad |\omega| \to \infty. \tag{4.3.27}$$

Also since

$$\lim_{\epsilon \to 0+} W(i + \epsilon e^{i\theta}) = \text{finite}$$

(although both D and $\zeta_1\zeta_3$ vanish at $\omega = i$) and

$$\lim_{\epsilon \to 0+} W(i/\sqrt{\beta} + \epsilon e^{i\theta}) = \text{finite}$$

it follows that

$$\int_{c_3} \to 0 \quad \text{and} \quad \int_{c_4} \to 0 \quad \text{as} \quad \epsilon \to 0+. \tag{4.3.28}$$

Then

$$\text{Re} \int_0^\infty W(\omega_1)d\omega_1 = \text{Re} \, P \int_1^\infty W(i\omega_2)id\omega_2$$

$$+ \, \text{Re} \, \pi i \, \{\text{Residue } W(\omega) \text{ at } \omega = i\omega_R\}$$

$$+ \, \text{Re} \int_{1/\sqrt{\beta}}^1 W(i\omega_2)id\omega_2 + \text{Re} \int_0^{1/\sqrt{\beta}} W(i\omega_2)id\omega_2.$$

$$(4.3.29)$$

But from (4.3.16) and (4.3.17), $\zeta_1(i\omega_2)$ and $\zeta_3(i\omega_2)$ are real for $0 \leqslant \omega_2 \leqslant 1/\sqrt{\beta}$ so that $W(i\omega_2)$ is real, hence

$$\text{Re} \int_0^{1/\sqrt{\beta}} W(i\omega_2)id\omega_2 = 0. \qquad (4.3.30)$$

Also from (4.3.16) and (4.3.17) $\zeta_1(i\omega_R)$ and $\zeta_3(i\omega_R)$ are both imaginary. This, together with the fact that $D(\omega)$ is a function of ω^2, implies

$$\text{Re} \, \pi i \, [\text{Residue } W(\omega) \text{ at } \omega = i\omega_R]$$

$$= \, \text{Re} \, \pi i \left[\frac{\zeta_1(i\omega_R)\zeta_3(i\omega_R) \, \{\zeta_1(i\omega_R) + \zeta_3(i\omega_R)\}}{D'(i\omega_R)} e^{-s\omega_R|y|} \right]$$

$$= \, \text{Re} \, \pi i \, [\text{real function of } \omega_R] = 0. \qquad (4.3.31)$$

Thus from (4.2.31) and above

$$\bar{w}(y, 0, s) = -\frac{P_0\alpha}{\sqrt{c_{44}\rho\pi}} \left[\text{Re} \, P \int_1^\infty \tilde{W}(i\omega_2)e^{-s\omega_2|y|}id\omega_2 \right.$$

$$\left. + \, \text{Re} \int_{1/\sqrt{\beta}}^1 \tilde{W}(i\omega_2)e^{-s\omega_2|y|}id\omega_2 \right]. \qquad (4.3.32)$$

But from the Laplace inversion integral

$$w(y, 0, \tau) = \frac{1}{2\pi i} \int_c \bar{w}(y, 0, s)e^{s\tau}ds,$$

and the relation (in the sense of distributions)

$$\frac{1}{2\pi i} \int_c e^{s(\tau - \omega_2|y|)} ds = \delta(\tau - \omega_2|y|) = \frac{1}{|y|}\delta(\omega_2 - T)$$

where

$$T = \tau/|y|, \tag{4.3.33}$$

the expression for w becomes

$$w(y, 0, \tau) = -\frac{P_0 \alpha}{\pi |y| \sqrt{c_{44}\rho}} \left[\operatorname{Re} P \int_1^\infty \tilde{W}(i\omega_2) \delta(\omega_2 - T) i d\omega_2 \right.$$

$$\left. + \operatorname{Re} \int_{1/\sqrt{\beta}}^1 \tilde{W}(i\omega_2) \delta(\omega_2 - T) i d\omega_2 \right]. \tag{4.3.34}$$

Now from the sifting property of the Dirac delta

$$\int_a^b f(\omega_2) \delta(\omega_2 - T) d\omega_2 = \begin{cases} f(T) & a < T < b \\ 0 & \text{otherwise} \end{cases}$$

$$= f(T) [H(T - a) - H(T - b)] \tag{4.3.35}$$

So

$$w(y, 0, \tau) = -\frac{P_0 \alpha}{\pi |y| \sqrt{c_{44}\rho}} \{ \operatorname{Re} i\tilde{W}(iT) [H(T - 1)$$

$$+ \operatorname{Re} i\tilde{W}(iT) [H(T - 1/\sqrt{\beta}) - H(T - 1)] \}. \tag{4.3.36}$$

Note that the Cauchy principle value P is no longer needed (it would be however if a further integration were required). The two terms within the brackets $\{\ldots\}$ in (4.3.36) may be combined to give the compact formula

$$w(y, 0, \tau) = -\frac{P_0 \alpha}{\pi |y| \sqrt{c_{44}\rho}} \operatorname{Re} i\tilde{W}(iT) H(T - 1/\sqrt{\beta}), \tag{4.3.37}$$

but this is deceptively simple, since there will be a change in $\operatorname{Re} i\tilde{W}(iT)$ as T passes through the value 1. Employing now the expressions (4.3.18)–(4.3.21) for $\zeta_1(i\omega_2)$ and $\zeta_3(i\omega_2)$ gives the explicit formula for the surface normal displacement

$$\frac{\pi |y| \sqrt{c_{44}\rho}}{P_0} w(y, 0, \tau) \equiv F(T) \tag{4.3.38}$$

as

$$0 \leqslant T \leqslant 1/\sqrt{\beta} \qquad F(T) = 0 \tag{4.3.39}$$

$$1/\sqrt{\beta} \leqslant T \leqslant 1 \qquad F(T) = -\frac{\alpha i \zeta_3(iT) \zeta_1^2(iT)}{A_1} B_1 \tag{4.3.40}$$

where

$$B_1 = \{2(1 - \kappa)(-T^2)(1 - T^2) - (-\gamma T^2 + \alpha)(1 - T^2) + \alpha \zeta_3^2(iT)\}, \tag{4.3.41}$$

$$A_1 = [2(1-\kappa)(-T^2)(1-T^2)-(-\gamma T^2 + \alpha)(1-T^2)]^2$$

$$+ \alpha(\beta T^2 - 1)(1 - T^2), \qquad (4.3.42)$$

and

$$1 \leqslant T < \infty \qquad F(T) = -\frac{\alpha \zeta_1(iT)\zeta_3(iT)\,[i\zeta_1(iT) + i\zeta_3(iT)]}{D(iT)}. \qquad (4.3.43)$$

Note that w is continuous at the wave fronts $T = 1/\sqrt{\beta}$ ($\tau = |y|/\sqrt{\beta}$) and $T = 1$ ($\tau = |y|$) where

$$\lim_{T \to 1/\sqrt{\beta}} F(T) = 0 \qquad (4.3.44)$$

and

$$\lim_{T \to 1} F(T) = -\left[\frac{\gamma - (\alpha + 1)}{\alpha}\right]^{1/2}. \qquad (4.3.45)$$

In the range $1 \leqslant T < \infty$, w has a simple pole at $T = \omega_R$ caused by the zero in the denominator of (4.3.43). As $T \to \infty$,

$$F(T) \sim \sqrt{\frac{\beta}{2}} \frac{[\gamma - \sqrt{\gamma^2 - 4\alpha\beta}\,]^{1/2} + [\gamma + \sqrt{\gamma^2 - 4\alpha\beta}\,]^{1/2}}{2(\kappa - 1) + \gamma} T^{-1}. \qquad (4.3.46)$$

It will now be shown that $F(T)$ has a double zero at $T = T_0$ where

$$T_0 = [(\kappa - 1 + \alpha)/(2\kappa - 2 + \gamma)]^{1/2} \qquad (4.3.47)$$

and that this point, which lies in the interval $1/\sqrt{\beta} < T < 1$ is a (local) maximum of $F(T)$. From (4.3.40) write

$$F(T) = \tfrac{1}{2}f(T)\phi(x) \quad \text{where} \quad f(T) = -\frac{\alpha i \zeta_3 \zeta_1^2}{A_1}$$

and

$$\phi(x) = 2B_1 = (1 - 2x)S + \kappa - [S^2 + 2\kappa(1 - 2x)S + \kappa^2]^{1/2}$$

where

$$S(x) = -2(1 - \kappa)x + \gamma x - \alpha + 1 - \kappa,$$

having replaced T^2 by x. Noting now that $S(x_0) = 0$ where $x_0 = T_0^2$, then it is easily shown that $\phi(x_0) = \phi'(x_0) = 0$ and that $\phi''(x_0) < 0$. Hence $F(T_0) = 0$ since $\phi(x_0) = 0$, $F'(T_0) = 0$ since $\phi(x_0) = \phi'(x_0) = 0$, and $F''(T_0) = \tfrac{1}{2}f(T_0)\phi''(x_0)(dx/dT)^2 < 0$ since $f(T_0) > 0$. Thus $F(T)$ vanishes at its local maximum point $T = T_0$.

Before turning to the inversion of the tangential surface displacement $v(y, 0, \tau)$, the isotropic version of equations (4.3.40) and (4.3.43) will be listed. Recall that for an isotropic solid $\alpha = \beta = \gamma/2$ so

$$1/\sqrt{\alpha} \leqslant T \leqslant 1, F(T) = -\frac{(2T^2 - 1)^2\sqrt{T^2 - 1/\alpha}}{\{(1 - 2T^2)^4 + 16T^4(T^2 - 1/\alpha)(1 - T^2)\}} \qquad (4.3.48)$$

and

$$1 \leqslant T \leqslant \infty, F(T) = -\frac{\sqrt{T^2 - 1/\alpha}}{(-2T^2 + 1)^2 - 4T^2\sqrt{T^2 - 1}\sqrt{T^2 - 1/\alpha}}.$$

$$(4.3.49)$$

The expressions (4.3.48) and (4.3.49) and their tangential displacement counterpart (to be given below) were the major achievements of Lamb's paper [57]. In the isotropic case, the double zero of $F(T)$ at $T_0 = 1/\sqrt{2}$ is obvious, and in agreement with (4.3.47) since $\kappa = \alpha - 1$ now.

v inversion

Call

$$V(\omega) = \tilde{V}(\omega)e^{is\omega|y|}, \qquad (4.3.50)$$

where

$$\tilde{V}(\omega) = \frac{\omega[\alpha \varsigma_1(\omega)\varsigma_3(\omega) + (1 - \kappa)(\omega^2 + 1)]}{D(\omega)} \qquad (4.3.51)$$

so that $V(\omega)$ is the integrand of (4.2.32) to within a multiplicative factor that does not depend on ω. Please observe that $V(\omega)$, when regarded as a function of the complex variable ω, has singularities only at the branch points $\omega = i$, $i/\sqrt{\beta}$ and the simple pole $\omega = i\omega_R$ in the upper-half ω-plane. Thus the restriction of this section to materials (1) of Table 11 notwithstanding, *the v inversion integral need only be evaluated once for all materials (1), (2) and (3) of Table 11.* This is not the case for the normal surface displacement due to the term $\varsigma_1(\omega) + \varsigma_2(\omega)$ which appears in (4.3.25).

   Following the integration technique used above on w, $V(\omega)$ will also be integrated around the contour shown in Figure 65. This leads to

$$\mathrm{Im} \int_0^\infty V(\omega_1)d\omega_1 = \mathrm{Im}\, P \int_1^\infty V(i\omega_2)id\omega_2$$

$$+ \mathrm{Im}\, i\{\mathrm{Residue}\, V(\omega)\, \text{at}\, \omega = i\omega_R\}$$

$$+ \mathrm{Im} \int_{1/\sqrt{\beta}}^1 V(i\omega_2)id\omega_2 + \mathrm{Im} \int_0^{1/\sqrt{\beta}} V(i\omega_2)id\omega_2.$$

$$(4.3.52)$$

But from (4.3.15) the product $\varsigma_1(i\omega_2)\varsigma_3(i\omega_2)$ is real for $\omega_2 \geqslant 1$ and for $0 \leqslant \omega_2 \leqslant 1/\sqrt{\beta}$ from (4.3.13). Consequently, from (4.3.51), $V(i\omega_2)$ is imaginary in these intervals so that

$$\mathrm{Im}\, P \int_1^\infty V(i\omega_2)id\omega_2 = 0 \quad \text{and} \quad \mathrm{Im} \int_1^{1/\sqrt{\beta}} V(i\omega_2)id\omega_2 = 0. \qquad (4.3.53)$$

Unlike the normal surface displacement, the residue from the Rayleigh pole does contribute to the tangential surface displacement.

$$\{\text{Residue } V(\omega) \text{ at } \omega = i\omega_R\} = \psi(\omega_R)e^{-s\omega_R|y|}, \tag{4.3.54}$$

where

$$\psi(\omega_R) = \frac{(1-\omega_R^2)[-2(1-\kappa)\omega_R^2+\gamma\omega_R^2-\alpha+1-\kappa]}{\left| 4(1-\kappa)(-2\omega_R^2+1)-2(-2\gamma\omega_R^2+\gamma+\alpha) \right.}$$
$$\left. -\frac{\alpha(\beta+1)-2\alpha\beta\omega_R^2}{(1-\omega_R^2)[-2(1-\kappa)\omega_R^2+\gamma\omega_R^2-\alpha]} \right\}. \tag{4.3.55}$$

Using the relation between v and $V(\omega)$ from (4.2.32) and performing the Laplace inversion of (4.3.52) gives,

$$v(y,0,\tau) = \frac{P_0 \text{ sgn } y}{|y|\sqrt{c_{44}\rho}} \psi(\omega_R)\delta(T-\omega_R)$$

$$+\frac{P_0 \text{ sgn } y}{\pi|y|\sqrt{c_{44}\rho}} \text{ Im } i\tilde{V}(iT) [H(T-1/\sqrt{\beta})-H(T-1)]. \tag{4.3.56}$$

Or, in terms of the dimensionless displacement function $G(T)$, (in analogy to the dimensionless normal displacement $F(T)$ defined in (4.3.38)) more explicitly

$$\frac{\pi|y|\sqrt{c_{44}\rho}}{P_0 \text{ sgn } y} v(y\ 0,\tau) \equiv G(T). \tag{4.3.57}$$

$$G(T) = \pi\psi(\omega_R)\delta(T-\omega_R)$$

$$-\frac{T\{-2(1-\kappa)T^2+\gamma T^2-\alpha+1-\kappa\}\sqrt{\alpha(\beta T^2-1)(1-T^2)}}{A_2}$$

$$\times [H(T-1/\sqrt{\beta})-H(T-1)] \tag{4.3.58}$$

where

$$A_2 = [-2(1-\kappa)T^2+\gamma T^2-\alpha]^2(1-T^2)+\alpha(\beta T^2-1). \tag{4.3.59}$$

Note from (4.3.42) that $A_1 = (1-T^2)A_2$. The displacement function $G(T)$ is continuous (but not its derivative) at the wave fronts $T = 1/\sqrt{\beta}$ and $T = 1$, where it vanishes. Also $G(T)$ has a (simple) zero at $T = T_0$ where T_0 was defined in (4.3.47).

In the isotropic situation, (4.3.58) yields

$$1/\sqrt{\alpha} \leqslant T \leqslant 1, G(T) = -\frac{2T(2T^2-1)\sqrt{(T^2-1/\alpha)(1-T^2)}}{\{(1-2T^2)^4+16T^4(T^2-1/\alpha)(1-T^2)\}}, \tag{4.3.60}$$

this expression being first established by Lamb [57].

This concludes the analysis of the surface displacements for all materials listed in Table 12. Some further comments and a graph will be given in section 4.6. below.

4. Transform Inversion for Materials Satisfying Condition (2) of Table 11

When the elastic parameter γ satisfies conditions (2) or (3) of Table 11, then the branch points of $\zeta(\omega)$ caused by the vanishing of $\phi(\omega)$ are no longer off the Fourier and Cagniard-Perkins integration paths. In fact $\zeta(\omega)$ will have branch points at

$$\omega = \pm\omega_3 \quad \text{and} \quad \omega = \pm i\omega_4 \tag{4.4.1}$$

where

$$\omega_3 = \sqrt{\omega_+}, \qquad \omega_4 = \sqrt{|\omega_-|} \tag{4.4.2}$$

and

$$\omega_\pm = \frac{2\alpha(\beta+1) - \gamma(\alpha+1) \mp 2[\alpha(\alpha+\beta-\gamma)(1+\alpha\beta-\gamma)]^{1/2}}{\gamma^2 - 4\alpha\beta} \tag{4.4.3}$$

Numerical values of these points, together with the Rayleigh pole location, for the four remaining crystals of Table 11 are listed in Table 13.

Table 13.   Branch points of $\zeta(\omega)$ and Rayleigh pole location in the first quadrant of the ω-plane for the (2) and (3) crystals of Table 11.

Crystal	α	β	γ	i	$i/\sqrt{\beta}$	ω_3	$i\omega_4$	$i\omega_R$	Remark
Apatite	2.11	2.52	2.34	i	$i0.62994$	0.27456	$i1.01731$	$i1.16590$	(3)
Beryllium	2.07	1.80	3.56	i	$i0.74536$	0.69172	$i1.03577$	$i1.15883$	(2)
Cadmium	2.62	5.95	6.80	i	$i0.40996$	0.32222	$i1.25238$	$i1.07245$	(3)
Zinc	1.57	4.17	2.40	i	$i0.48970$	0.12603	$i1.00070$	$i1.13860$	(3)

(2) $(\beta+1) < \gamma < (\alpha+\beta)$ and $(\gamma^2 - 4\alpha\beta) < 0$.
(3) $\gamma < (\beta+1)$, $(\gamma^2 - 4\alpha\beta) < 0$ and $\beta > \alpha$.

The branch cuts for $\zeta_1(\omega)$ and $\zeta_3(\omega)$ will now be defined. Since

$$\phi(\omega) = (\gamma^2 - 4\alpha\beta)(\omega^2 - \omega_3^2)(\omega^2 + \omega_4^2) \tag{4.4.4}$$

then define

$$\sqrt{\phi(\omega)} = \sqrt{4\alpha\beta - \gamma^2}\, e^{-i\pi/2} (r_1 r_2 r_3 r_4)^{1/2}\, e^{i(\theta_1+\theta_2+\theta_3+\theta_4)/2} \tag{4.4.5}$$

where

$$\begin{aligned}
\omega - \omega_3 &= r_1 e^{i\theta_1} & \text{with} \quad r_1 &= |\omega - \omega_3|, & -\pi < \theta_1 < \pi \\
\omega + \omega_4 &= r_2 e^{i\theta_2} & r_2 &= |\omega + \omega_3|, & -\pi < \theta_2 < \pi \\
\omega + i\omega_4 &= r_3 e^{i\theta_3} & r_3 &= |\omega + i\omega_4|, & -\pi < \theta_3 < \pi \\
\omega - i\omega_4 &= r_4 e^{i\theta_4} & r_4 &= |\omega - i\omega_4|. & -\pi < \theta_4 < \pi
\end{aligned} \tag{4.4.6}$$

With the above definition of $\sqrt{\phi(\omega)}$, this function on $\omega = \omega_1 + i0$ is positive imaginary for $\omega_1 < -\omega_3$, positive for $-\omega_3 < \omega_1 < \omega_3$, and negative imaginary for $\omega_1 > \omega_3$. The functions $\zeta_1(\omega)$ and $\zeta_3(\omega)$ may now be rendered single valued on $\omega = \omega_1 + i0$ by setting

$$\zeta_1^2 = a + ib \quad \text{and} \quad \zeta_1 = \sqrt{a^2 + b^2}\, e^{i\phi_1/2} \tag{4.4.7}$$

where $\quad \tan \phi_1 = b/a \quad$ and $\quad -\pi < \phi_1 < \pi,$

$$\zeta_3^2 = c + id \quad \text{and} \quad \zeta_3 = \sqrt{c^2 + d^2}\, e^{i\phi_3/2} \tag{4.4.8}$$

where $\quad \tan \phi_3 = c/d \quad$ and $\quad -\pi < \phi_3 < \pi.$

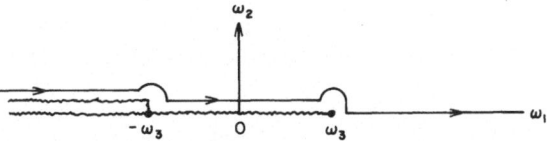

Figure 66.   Real Fourier Inversion Path for Materials (2) and (3) of Table 11.

With the Fourier inversion path indented as indicated in Figure 66 and the functions $\zeta_1(\omega)$ and $\zeta_3(\omega)$ defined on this path by the help of formulas (4.4.7) and (4.4.8) it follows that

$$\text{Re}\,\zeta_1 > 0 \quad \text{and} \quad \text{Re}\,\zeta_3 > 0 \quad \text{on} \quad \omega = \omega_1 + i0. \tag{4.4.9}$$

Consequently the analysis of section 4.2 remains valid for materials (2) and (3), the crucial point being the elimination of the exponential terms $e^{s\zeta_1 z}$ and $e^{s\zeta_3 z}$ due to the boundary condition at $z = \infty$. In this connection it should also be noted that on $\omega = \omega_1 + i0$, $\text{Im}(\zeta_1 + \zeta_3) = 0$, i.e. $(\zeta_1 + \zeta_3) > 0$ on the integration path. This follows from (4.2.14), (4.4.7) and (4.4.8).

In this section, attention will be concentrated on materials satisfying condition (2) of Table 11, of which Beryllium is the solitary example. For Beryllium, from (4.2.14) $\zeta_1(i) = 0, \zeta_3(i) \neq 0, \zeta_1(i/\sqrt{\beta}) \neq 0$ and $\zeta_3(i/\sqrt{\beta}) = 0$. Hence

$\omega = i$	(and $\omega = -i$)	is a branch point of $\zeta_1(\omega)$,	(4.4.10)
$\omega = i$	(and $\omega = -i$)	is not a branch point of $\zeta_3(\omega)$,	(4.4.11)
$\omega = i/\sqrt{\beta}$	(and $\omega = -i/\sqrt{\beta}$)	is not a branch point of $\zeta_1(\omega)$,	(4.4.12)
$\omega = i/\sqrt{\beta}$	(and $\omega = -i/\sqrt{\beta}$)	is a branch point of $\zeta_3(\omega)$.	(4.4.13)

In view of the definition of $\sqrt{\phi(\omega)}$ from (4.4.5) it follows that on the imaginary ω-axis

$$\sqrt{\phi(i\omega_2)} = i\sqrt{(4\alpha\beta - \gamma^2)(\omega_2^2 + \omega_3^2)(\omega_2^2 - \omega_4^2)} \quad \text{for} \quad \omega_2 \geqslant \omega_4, \tag{4.4.14}$$

and

$$\sqrt{\phi(i\omega_2)} = \sqrt{(4\alpha\beta - \gamma^2)(\omega_2^2 + \omega_3^2)(\omega_4^2 - \omega_2^2)} \quad \text{for} \quad 0 \leqslant \omega_2 \leqslant \omega_4. \tag{4.4.15}$$

From a reference to (4.2.14) and the discussion preceeding (4.4.10) it follows that

$$\zeta_1^2(i\omega_2) > 0 \quad \text{for} \quad 0 \leqslant \omega_2 < 1 \quad \text{and} \quad \zeta_1^2(i\omega_2) < 0 \quad \text{for} \quad 1 < \omega_2 < \omega_4 \tag{4.4.16}$$

and

$$\zeta_3^2(i\omega_2) > 0 \quad \text{for} \quad 0 \leqslant \omega_2 < 1/\sqrt{\beta}, \qquad \zeta_3^2(i\omega_2) < 0 \quad \text{for}$$

$$1/\sqrt{\beta} < \omega_2 < \omega_4. \tag{4.4.17}$$

Of course both $\zeta_1^2(i\omega_2)$ and $\zeta_3^2(i\omega_2)$ are complex for $\omega_2 > \omega_4$ by virtue of (4.4.14).

In view of the above behaviour of $\zeta_1^2(i\omega_2)$ and $\zeta_3^2(i\omega_2)$ and the desired form for the product $\zeta_1(i\omega_2)\zeta_3(i\omega_2)$ as given in equations (4.3.13)–(4.3.15) it is consistent to take

$$\zeta_1(i\omega_2) = i\{\gamma\omega_2^2 - (\alpha + 1) - i\sqrt{R_1(\omega_2)}\}^{1/2}/\sqrt{2\alpha}, \qquad \omega_4 \leqslant \omega_2 < \infty$$

where
$$\tag{4.4.18}$$

$$R_1(\omega_2) = (4\alpha\beta - \gamma^2)(\omega_2^2 + \omega_3^2)(\omega_2^2 - \omega_4^2), \tag{4.4.19}$$

$$\zeta_1(i\omega_2) = i\{\gamma\omega_2^2 - (\alpha + 1) - \sqrt{R_2(\omega_2)}\}^{1/2}/\sqrt{2\alpha}, \qquad 1 \leqslant \omega_2 \leqslant \omega_4$$

$$\tag{4.4.20}$$

where

$$R_2(\omega_2) = (4\alpha\beta - \gamma^2)(\omega_2^2 + \omega_3^2)(\omega_4^2 - \omega_2^2), \tag{4.4.21}$$

$$\zeta_1(i\omega_2) = \{-\gamma\omega_2^2 + (\alpha + 1) + \sqrt{R_2(\omega_2)}\}^{1/2}/\sqrt{2\alpha}, \qquad 0 \leqslant \omega_2 \leqslant 1. \tag{4.4.22}$$

Also

$$\zeta_3(i\omega_2) = i\{\gamma\omega_2^2 - (\alpha + 1) + i\sqrt{R_1(\omega_2)}\}^{1/2}/\sqrt{2\alpha}, \qquad \omega_4 \leqslant \omega_2 < \infty$$

$$\tag{4.4.23}$$

$$\zeta_3(i\omega_2) = i\{\gamma\omega_2^2 - (\alpha + 1) + \sqrt{R_2(\omega_2)}\}^{1/2}/\sqrt{2\alpha}, \qquad 1/\sqrt{\beta} \leqslant \omega_2 \leqslant \omega_4$$

$$\tag{4.4.24}$$

$$\zeta_3(i\omega_2) = \{-\gamma\omega_2^2 + (\alpha + 1) - \sqrt{R_2(\omega_2)}\}^{1/2}/\sqrt{2\alpha}, \qquad 0 \leqslant \omega_2 \leqslant 1/\sqrt{\beta}. \tag{4.4.25}$$

The above expressions will cause the product $\zeta_1(i\omega_2)\zeta_3(i\omega_2)$ to conform to its previous forms as given in (4.3.13)–(4.3.15). Since it is needed in the evaluation of the surface normal displacement, the sum $\zeta_1(i\omega_2) + \zeta_3(i\omega_2)$ will also be recorded here

$$\zeta_1(i\omega_2) + \zeta_3(i\omega_2) = i2\text{Re}\{\gamma\omega_2^2 - (\alpha + 1) + i\sqrt{R_1(\omega_2)}\}^{1/2}/\sqrt{2\alpha}$$

$$\text{for} \quad \omega_4 \leqslant \omega_3 < \infty, \tag{4.4.26}$$

$$\zeta_1(i\omega_2) + \zeta_3(i\omega_2) = i\{\gamma\omega_2^2 - (\alpha + 1) - \sqrt{R_2(\omega_2)}\}^{1/2}/\sqrt{2\alpha}$$

$$+ i\{\gamma\omega_2^2 - (\alpha + 1) + \sqrt{R_2(\omega_2)}\}^{1/2}/\sqrt{2\alpha}$$

$$\text{for} \quad 1 \leqslant \omega_2 \leqslant \omega_4, \tag{4.4.27}$$

$$\zeta_1(i\omega_2) + \zeta_3(i\omega_2) = \{-\gamma\omega_2^2 + (\alpha+1) + \sqrt{R_2(\omega_2)}\}^{1/2}/\sqrt{2\alpha}$$

$$+ i\{\gamma\omega_2^2 - (\alpha+1) + \sqrt{R_2(\omega_2)}\}^{1/2}/\sqrt{2\alpha}$$

$$\text{for} \quad 1/\sqrt{\beta} \leqslant \omega_2 \leqslant 1, \tag{4.4.28}$$

and

$$\zeta_1(i\omega_2) + \zeta_3(i\omega_2) = \{-\gamma\omega_2^2 + (\alpha+1) + \sqrt{R_2(\omega_2)}\}^{1/2}/\sqrt{2\alpha}$$

$$+ \{-\gamma\omega_2^2 + (\alpha+1) - \sqrt{R_2(\omega_2)}\}^{1/2}/\sqrt{2\alpha}$$

$$\text{for} \quad 0 \leqslant \omega_2 \leqslant 1/\sqrt{\beta}. \tag{4.4.29}$$

w inversion

The expression $W(\omega)$, as given in (4.3.24) will now be integrated around the closed contour shown in figure 67.

$$\int_0^\infty W(\omega_1)d\omega_1 + \int_{c_1} + P\int_\infty^{\omega_4} W(i\omega_2)id\omega_2 - \pi i \{\text{Residue } W(\omega) \text{ at } \omega = i\omega_R\}$$

$$+ \int_{c_5} + \int_{\omega_4}^1 W(i\omega_2)id\omega_2 + \int_{c_3} + \int_1^{1/\sqrt{\beta}} W(i\omega_2)id\omega_2$$

$$+ \int_{c_4} + \int_{1/\sqrt{\beta}}^0 W(i\omega_2)id\omega_2 = 0. \tag{4.4.30}$$

By a treatment similar to that used in section 4.3, the indentation integrals along c_3, c_4 and c_5 and the loop integral along c_1 can be shown to vanish when a suitable limit·

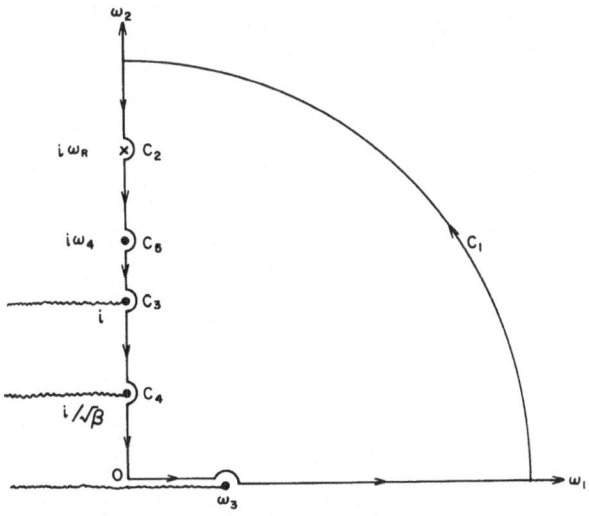

Figure 67.   Complex ω-Plane Showing the Contours used in Evaluating the Surface Displacement w for (2) and (3) Materials of Table 11.

is taken. Then

$$\text{Re} \int_0^\infty W(\omega_1)d\omega_1 \;=\; \text{Re} \, P \int_{\omega_4}^\infty W(i\omega_2)id\omega_2 + \text{Re} \, \pi i \, \{\text{Residue } W(\omega) \text{ at}$$

$$\omega = i\omega_R\}$$

$$+ \, \text{Re} \int_1^{\omega_4} W(i\omega_2)id\omega_2 + \text{Re} \int_{1/\sqrt{\beta}}^1 W(i\omega_2)id\omega_2$$

$$+ \, \text{Re} \int_0^{1/\sqrt{\beta}} W(i\omega_2)id\omega_2. \tag{4.4.31}$$

But for $0 \leqslant \omega_2 \leqslant 1/\sqrt{\beta}$, $W(i\omega_2) = $ real so

$$\text{Re} \int_0^{1/\sqrt{\beta}} W(i\omega_2)id\omega_2 \;=\; 0. \tag{4.4.32}$$

Also $\{\text{Residue } W(\omega) \text{ at } \omega = i\omega_R\} = $ real function of ω_R, so that the residue contribution vanishes. Following now the w inversion of section (4.3) gives

$$F(T) \;=\; -\alpha \, \text{Re} \, i\widetilde{W}(iT)H(T - 1/\sqrt{\beta}) \tag{4.4.33}$$

where $F(T)$ is the dimensionless normal surface displacement defined by (4.3.38) and T was previously given in (4.3.33). More explicitly (4.4.33) becomes

$$0 \leqslant T \leqslant 1/\sqrt{\beta}, \qquad F(T) \;=\; 0, \tag{4.4.34}$$

$$1/\sqrt{\beta} \leqslant T \leqslant 1, \qquad F(T) \;=\; -\frac{\alpha i \zeta_3(iT)\zeta_1^2(iT)}{A_1} B_1 \tag{4.4.35}$$

$$1 \leqslant T \leqslant \omega_4, \qquad F(T) \;=\; -\sqrt{\frac{\beta}{2}} \, \frac{\sqrt{(T^2 - 1/\beta)(T^2 - 1)}}{D(iT)} B_2, \tag{4.4.36}$$

where

$$B_2 \;=\; [\{\gamma T^2 - \alpha - 1 - \sqrt{R_2(T)}\}^{1/2} + \{\gamma T^2 - \alpha - 1 + \sqrt{R_2(T)}\}^{1/2}], \tag{4.4.37}$$

$$\omega_4 \leqslant T < \infty, \qquad F(T) \;=\; -\frac{\sqrt{\beta}\sqrt{(T^2 - 1/\beta)(T^2 - 1)}}{D(iT)} B_3, \tag{4.4.38}$$

where

$$B_3 \;=\; \{\gamma T^2 - \alpha - 1 + 2\sqrt{\alpha\beta}\sqrt{(T^2 - 1/\beta)(T^2 - 1)}\}^{1/2}. \tag{4.4.39}$$

Expressions for B_1 and A_1 have been given in (4.3.41) and (4.3.42) respectively. Thus equations (4.4.34)–(4.4.39) together with the values of ζ_1 and ζ_3 on $\omega = i\omega_2$ from this section, comprise an explicit closed form solution for the normal surface displacement. Again w is continuous at the wave fronts $T = 1/\sqrt{\beta}$ and $T = 1$ where

$$\lim_{T \to 1/\sqrt{\beta}} F(T) = 0 \quad \text{and} \tag{4.4.40}$$

$$\lim_{T \to 1} F(T) = -\left[\frac{\gamma - (\alpha + 1)}{\alpha}\right]^{1/2}. \tag{4.4.41}$$

Also w is continuous and in fact analytic at $T = \omega_4$. It is a curious feature of Beryllium that an apparent wave front should be introduced at $T = \omega_4$. Again $F(T)$ contains a simple pole at $T = \omega_R$. As $T \to \infty$

$$F(T) \sim \frac{\sqrt{\beta}\{\gamma + 2\sqrt{\alpha\beta}\}^{1/2}}{2(\kappa - 1) + \gamma} T^{-1} \tag{4.4.42}$$

In the interval $1/\sqrt{\beta} < T < 1$, $F(T)$ again contains a double zero at $T = T_0$ just as in the previous section.

v-inversion

Since the (4.2.32) expression for $\bar{v}(y, 0, s)$ encounters $\zeta_1(\omega)$ and $\zeta_3(\omega)$ only through the product $\zeta_1(\omega)\zeta_3(\omega)$ and the fact that this product is treated precisely the same way for (1) and (2) materials in Table 11, it follows that the (4.3.58) formula for $G(T)$ remains valid for this section also.

Thus the surface displacement analysis for (2) materials is now complete. Due to the points $\omega = \omega_3$ and $\omega = i\omega_4$ the analysis of this section was somewhat more complicated than that given for the (1) materials in section 4.3 and yet the final results will graph very much the same. It turns out that Beryllium, in some respects, acts as a transition material leading from the isotropic-like behaviour of materials (1) to the new wave front behaviour of materials (3), in particular Apatite and Zinc.

5. Transform Inversion for Materials Satisfying Condition (3) of Table 11

There are three crystals in Table 11 whose surface displacements have yet to be discussed, Apatite, Cadmium and Zinc. The crystal parameters for these materials satisfy the inequalities

$$\gamma < (\beta + 1), (\gamma^2 - 4\alpha\beta) < 0 \quad \text{with} \quad \beta > \alpha. \tag{4.5.1}$$

Of these three crystals, the analysis for Cadmium is most readily disposed of. In fact, the formulas for $\zeta_1(i\omega_2)$ and $\zeta_3(i\omega_2)$ are identical with those for Beryllium given in the previous section. For this reason the Beryllium surface displacement formulas, (4.4.34)–(4.4.39) for $F(T)$ and (4.3.58) for $G(T)$, are also valid for Cadmium. There is just one change, which is evident from Table 13, in that $\omega_4 > \omega_R$ for Cadmium. This, of course, requires a suitable modification of the Cagniard-Perkeris integration path shown in Figure 67. The fact that $\omega_4 > \omega_R$ implies that $F(T)$ will have a simple pole (at $T = \omega_R$) in the T interval of formula (4.4.36) rather than the T interval of formula (4.4.38) as in the case for Beryllium.

In the case of Apatite and Zinc (the analysis of the surface displacements is identical for both crystals) there will be branch points, for Im $\omega \geqslant 0$, located at

$$-\omega_3, \omega_3, i\omega_4 \quad \text{for} \quad \zeta_1(\omega) \tag{4.5.2}$$

and

$$-\omega_3, \omega_3, i\omega_4, i, i/\sqrt{\beta} \quad \text{for} \quad \zeta_3(\omega). \tag{4.5.3}$$

Also

$$\zeta_1^2(i\omega_2) > 0 \quad \text{for} \quad 0 \leqslant \omega_2 \leqslant \omega_4 \tag{4.5.4}$$

while

$$\zeta_3^2(i\omega_2) > 0 \qquad 0 \leqslant \omega_2 < 1/\sqrt{\beta} \tag{4.5.5}$$

$$\zeta_3^2(i/\sqrt{\beta}) = 0 \tag{4.5.6}$$

$$\zeta_3^2(i\omega_2) < 0 \qquad 1/\sqrt{\beta} < \omega_2 < 1 \tag{4.5.7}$$

$$\zeta_3^2(i) = 0 \tag{4.5.8}$$

$$\zeta_3^2(i\omega_2) > 0, \qquad 1 < \omega_2 < \omega_4. \tag{4.5.9}$$

The fact that $\zeta_3^2(i\omega_2)$ changes sign twice in the interval $0 < \omega_2 < \omega_4$ accounts for a new feature (lacuna) not previously encountered in the normal surface displacements of the other eleven crystals. Numerical values for $1/\sqrt{\beta}$, ω_3, ω_4 and ω_R are given in Table 13.

Expressions for $\zeta_1(i\omega_2)$ and $\zeta_3(i\omega_2)$ consistent with the above are given by

$$\zeta_1(i\omega_2) = i\{\gamma\omega_2^2 - (\alpha+1) - i\sqrt{R_1(\omega_2)}\}^{1/2}/\sqrt{2\alpha}, \qquad \omega_4 \leqslant \omega_2 < \infty \tag{4.5.10}$$

note that $\gamma\omega_2^2 - (\alpha+1)$ changes sign in this interval, here use $-\pi \leqslant \arg\{\ldots\} < 0$,

$$\zeta_1(i\omega_2) = \{-\gamma\omega_2^2 + (\alpha+1) + \sqrt{R_2(\omega_2)}\}^{1/2}/\sqrt{2\alpha}, \qquad 0 \leqslant \omega_2 \leqslant \omega_4 \tag{4.5.11}$$

here $\{\ldots\} > 0$,

$$\zeta_3(i\omega_2) = i\{\gamma\omega_2^2 - (\alpha+1) + i\sqrt{R_1(\omega_2)}\}^{1/2}/\sqrt{2\alpha}, \qquad \omega_4 \leqslant \omega_2 < \infty \tag{4.5.12}$$

note that $\gamma\omega_2^2 - (\alpha+1)$ changes sign in this interval, here use $0 < \arg\{\ldots\} \leqslant \pi$,

$$\zeta_3(i\omega_2) = -\{-\gamma\omega_2^2 + (\alpha+1) - \sqrt{R_2(\omega_2)}\}^{1/2}/\sqrt{2\alpha}, \qquad 1 \leqslant \omega_2 \leqslant \omega_4 \tag{4.5.13}$$

here $\{\ldots\} \geqslant 0$ with $=$ iff $\omega_2 = 1$.

$$\zeta_3(i\omega_2) = i\{\gamma\omega_2^2 - (\alpha+1) + \sqrt{R_2(\omega_2)}\}^{1/2}/\sqrt{2\alpha}, \qquad 1/\sqrt{\beta} \leqslant \omega_2 \leqslant 1 \tag{4.5.14}$$

here $\{\ldots\} \geqslant 0$ with = iff $\omega_2 = 1/\sqrt{\beta}$ or $\omega_2 = 1$,

$$\zeta_3(i\omega_2) = \{-\gamma\omega_2^2 + (\alpha + 1) - \sqrt{R_2(\omega_2)}\}^{1/2}/\sqrt{2\alpha}, \qquad 0 \leqslant \omega_2 \leqslant 1/\sqrt{\beta}$$

(4.5.15)

here $\{\ldots\} \geqslant 0$ with = iff $\omega_2 = 1/\sqrt{\beta}$.

The above expressions will cause the product $\zeta_1(\omega_2)\zeta_3(i\omega_2)$ to conform to the previous formulas (4.3.13)–(4.3.15). From equations (4.5.10)–(4.5.15) the sum $\zeta_1(i\omega_2) + \zeta_3(i\omega_2)$ may be written down as

$$\zeta_1(i\omega_2) + \zeta_3(i\omega_2) = i2\mathrm{Re}\,\{\gamma\omega_2^2 - (\alpha + 1) + i\sqrt{R_1(\omega_2)}\}^{1/2}/\sqrt{2\alpha}$$

$$\omega_4 \leqslant \omega_2 < \infty \qquad\qquad (4.5.16)$$

here $0 < \arg\{\ldots\} \leqslant \pi$,

$$\zeta_1(i\omega_2) + \zeta_3(i\omega_2) = \{-\gamma\omega_2^2 + (\alpha + 1) + \sqrt{R_2(\omega_2)}\}^{1/2}/\sqrt{2\alpha}$$

$$-\{-\gamma\omega_2^2 + (\alpha + 1) - \sqrt{R_2(\omega_2)}\}^{1/2}/\sqrt{2\alpha},$$

$$1 \leqslant \omega_2 \leqslant \omega_4 \qquad\qquad (4.5.17)$$

$$\zeta_1(i\omega_2) + \zeta_3(i\omega_2) = \{-\gamma\omega_2^2 + (\alpha + 1) + \sqrt{R_2(\omega_2)}\}^{1/2}/\sqrt{2\alpha}$$

$$+ i\{\gamma\omega_2^2 - (\alpha + 1) + \sqrt{R_2(\omega_2)}\}^{1/2}/\sqrt{2\alpha},$$

$$1/\sqrt{\beta} \leqslant \omega_2 \leqslant 1 \qquad\qquad (4.5.18)$$

and

$$\zeta_1(i\omega_2) + \zeta_3(i\omega_2) = \{-\gamma\omega_2^2(\alpha + 1) + \sqrt{R_2(\omega_2)}\}^{1/2}/\sqrt{2\alpha}$$

$$+ \{-\gamma\omega_2^2 + (\alpha + 1) - \sqrt{R_2(\omega_2)}\}^{1/2}/\sqrt{2\alpha}.$$

$$0 \leqslant \omega_2 \leqslant 1/\sqrt{\beta} \qquad\qquad (4.5.19)$$

w inversion

Figure 67 shows the ω-plane integration contour. Following the w inversion given in section 4.4,

$$\mathrm{Re}\int_0^\infty W(\omega_1)d\omega_1 = \mathrm{Re}\,P\int_{\omega_4}^\infty W(i\omega_2)id\omega_2 + \mathrm{Re}\int_1^{\omega_4} W(i\omega_2)id\omega_2$$

$$+ \mathrm{Re}\int_{1/\sqrt{\beta}}^1 W(i\omega_2)id\omega_2 + \mathrm{Re}\int_0^{1/\sqrt{\beta}} W(i\omega_2)id\omega_2,$$

(4.5.20)

since the indentation contours c_3, c_4, c_5, the loop contour c_1 and the real part of the residue contribution all vanish. Now for $0 \leqslant \omega_2 \leqslant 1/\sqrt{\beta}$, $\zeta_1(i\omega_2)$ and $\zeta_3(i\omega_2)$ and

153

hence $W(i\omega_2)$ are real so that

$$\text{Re} \int_0^{1/\sqrt{\beta}} W(i\omega_2)i d\omega_2 = 0. \tag{4.5.21}$$

Also for $1 \leqslant \omega_2 \leqslant \omega_4, \zeta_1(i\omega_2), \zeta_3(i\omega_2)$ and $W(i\omega_2)$ are real, hence

$$\text{Re} \int_1^{\omega_4} W(i\omega_2)i d\omega_2 = 0 \quad \text{(lacuna)}. \tag{4.5.22}$$

Thus, upon performing the Laplace inversion,

$$F(T) = -\alpha\{\text{Re}\, i\tilde{W}(iT)[H(T-\omega_4)] + \text{Re}\, i\tilde{W}(iT)\,[H(T-1/\sqrt{\beta}\,)$$
$$-H(T-1)]\}. \tag{4.5.23}$$

Written out explicitly the dimensionless normal surface displacement becomes

$$0 \leqslant T \leqslant 1/\sqrt{\beta}, \qquad F(T) = 0, \tag{4.5.24}$$

$$1/\sqrt{\beta} \leqslant T \leqslant 1, \qquad F(T) = -\frac{\alpha i \zeta_3(iT)\zeta_1^2(iT)}{A_1} B_1, \tag{4.5.25}$$

$$1 \leqslant T \leqslant \omega_4, \qquad F(T) = 0, \tag{4.5.26}$$

$$\omega_4 \leqslant T < \infty, \qquad F(T) = -\frac{\sqrt{\beta}\,\sqrt{T^2 - 1/\beta}\,(T^2 - 1)}{D(iT)} B_3 \tag{4.5.27}$$

Expressions for A_1, B_1 and B_3 were previously given in (4.3.41), (4.3.42) and (4.4.39).

The surface normal displacement now has three wave fronts at $T = 1/\sqrt{\beta}$, $T = 1$ and $T = \omega_4$ where $F(T)$ is continuous (in fact zero) but has infinite slope. It is particularly interesting to note that neither edge of the lacuna carries a singularity in the displacement (of course it does in the strain) unlike the free space problem treated in section 3.5. The free space problem singularity is apparently moderated by the wave interaction at the stress free surface of the half-space. $F(T)$ also contains a double zero at $T = T_0$ (see equation (4.3.47)) in the interval $1/\sqrt{\beta} < T < 1$, a simple pole at $T = \omega_R$ in the interval $T > \omega_4$ and has the behaviour (4.4.42) for large T.

As in section 4.4, the (4.3.58) formula for the tangential displacement $G(T)$ remains valid here. This completes the surface displacement analysis for the crystals Apatite, Cadmium and Zinc which are characterized as the (3) materials in Table 11.

6. Graph of the Surface Displacements for Some Hexagonal Crystals

Some of the gross features of the surface displacements obtained in this chapter can be explained by a reference to the slowness curves for the corresponding free space (unbounded domain) problem treated in Chapter 2. Note that the p and q slowness axes correspond respectively to the z and y material axes. Slowness figures $1-5$ then show that the outer branch of a slowness curve N_+ is locally convex where this branch is pierced by the q-axis for all classes except III and IV. Thus, from Table 7, it might

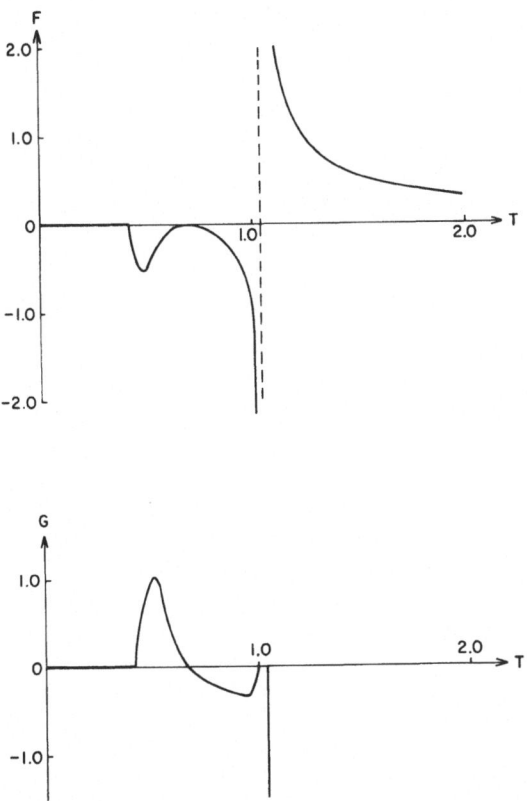

Figure 68.   Surface Displacements for Thallium Caused by an Impulsive Surface Line Load.

be expected that the crystals of classes I, II and V will exhibit common behaviour, at least so far as wave front arrivals are concerned. The results of this chapter have shown this to be true since the I, II and V class crystals have but two wave front arrivals $T = 1/\sqrt{\beta}$ and $T = 1$. On the other hand figure 4 shows that the N_+ branch of the slowness curve is not convex along the q-axis and the corresponding wave front curve (figure 14) reveals that three fronts are present. Again this is borne out in the half-space problem for the crystals Apatite and Zinc which have fronts $T = 1/\sqrt{\beta}$, $T = 1$ and $T = \omega_4$, the latter being related to the bitangent on N_+ parallel to the p-axis.

In order to illustrate the results of this chapter, graphs of the (dimensionless) surface displacements $F(T)$ (normal displacement) and $G(T)$ (tangential displacement), as a function of $T (= \tau/|y|)$, are shown in figures 68 and 69 for Thallium and Apatite. For the nine crystals obeying condition (1) of Table 11, the eight crystals Beryl, Cobalt, Hafnium, Ice, Magnesium, Rhenium, Titanium and Yttrium all have a first minimum on the F graph which is very close (within 0.01 or less) to the arrival at $T = 1/\sqrt{\beta}$. For this reason the graph of Thallium was selected to illustrate the (1) materials. Beryllium, the only (2) material in Table 11, has a graph with the same features as Thallium as, indeed, is the case for the (3) crystal Cadmium also. In the case of Beryllium the time (for fixed $|y|$) between the first arrival at $T = 0.74535599$ and arrival of the F minimum at $T = 0.74535699$ is extremely short. Note the Rayleigh

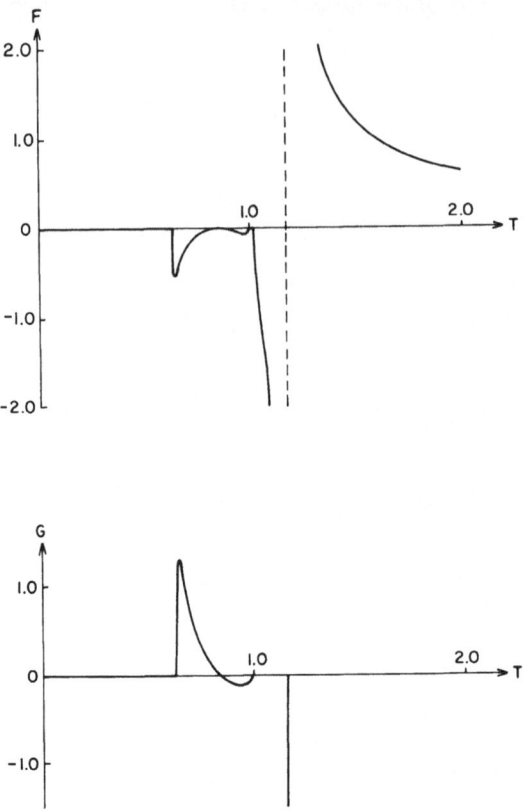

Figure 69.   Surface Displacements for Apatite Caused by an Impulsive Surface Line Load.

pole discontinuity (located by the dashed line) on the F graph and the vertical line on the G graph corresponding to the travelling (Dirac δ) surface wave. Also the zero of F and G at $T = T_0$ is clearly in evidence.

Figure 68 illustrates the graph of the remaining (3) materials, Apatite and Zinc. Apatite is shown here since its lacuna is a little wider than that of Zinc. The tangential displacements for both Thallium and Apatite behave in a similar way but there are significant differences in the normal displacements due mainly to the lacuna.

5

Epicenter and epicentral-axis motion of a three-dimensional half-space

Ideally, this chapter would be devoted to a three-dimensional treatment of Lamb's problem for a point load suddenly applied at the surface of a transversely isotropic elastic half-space. For an isotropic solid, the surface displacements for this problem were first successfully treated by Pekeris [48]. In the analysis of Pekeris, the Lamé parameters λ and μ of the solid were assumed to be equal. This assumption permitted an easy solution of the Rayleigh cubic equation, but was not actually crucial to the mathematical technique employed. Pekeris gave a closed form solution for the surface displacements when the applied point loading was normal to the half-space. Although both the normal and tangential components of the surface displacement vector were found explicitly, the tangential component (unlike the normal component) required elliptic integrals of the first and third kinds for its description. The corresponding problem for a transversely isotropic half-space, in which the surface is normal to the axis of material symmetry, has been studied by Ryan [67]. Ryan discusses the wave front shape and singularities in the interior of the solid and gives graphical results for the surface displacements. Unfortunately the three-dimensional version of Lamb's problem for a transversely isotropic solid does not (in contrast with the two-dimensional problem treated in Chapter 4) admit an explicit solution for surface displacements. Ryan uses numerical integration to evaluate the integral expressions for both components of the surface displacement vector.

In conformity with the general spirit of this monograph, which seeks to treat those aspects of transient wave propagation which admit explicit representation of the displacement field, attention in Chapter 5 will be directed toward the epicenter motion of the surface due to a buried point source and to the epicentral-axis motion caused by a normal point load which is suddenly applied on the surface. It will be shown that these two problems, which are intimately connected by the Betti-Rayleigh reciprocal theorem, can be treated explicitly although the results are far from trivial. Finally the chapter will end with a brief treatment of the body forces equivalent to internal discontinuities in an anisotropic elastic solid. This important work of Burridge and Knopoff [68] has application to problems in seismology and dislocations in crystals.

1. Problem Formulation for the Epicenter Motion of a Half-Space Due to the Sudden Application of a Buried Point Source

The half-space $-\infty < x, y < \infty, z > 0$ is occupied by a transversely isotropic solid, in which the z-axis coincides with the axis of material symmetry. At time $t = 0$, the solid is excited by a point body force $f_i = (a_1, a_2, a_3)\delta(x)\delta(y)\delta(z-h)H(t)$ located at a depth $h > 0$ below the surface on the z-axis. The body force has constant direction (a_1, a_2, a_3) and a (unit) step function time dependence. The surface $z = 0$ is free of stress, while prior to $t = 0$, the solid is at rest. The problem is to find the subsequent motion of the half-space.

An integral transform solution to the problem will be obtained by applying an exponential Fourier transform in the x and y directions, a Laplace transform in time and then solving directly the resulting set of ordinary differential equations in the variable z. If u, v and w are the components of the displacement vector, then the transformed equations of motion are (see equations (1.5.10), (1.5.11) together with (1.5.22) and (1.5.23))

$$\left(-\beta\xi^2 - \delta\eta^2 + \frac{\partial^2}{\partial z^2} - s^2\right)\bar{u}^* - (\beta-\delta)\xi\eta\bar{v}^* + i\kappa\xi\frac{\partial\bar{w}^*}{\partial z} = -\frac{\rho a_1}{c_{44}s}\delta(z-h),$$

$$-(\beta-\delta)\xi\eta\bar{u}^* + \left(-\delta\xi^2 - \beta\eta^2 + \frac{\partial^2}{\partial z^2} - s^2\right)\bar{v}^* + i\kappa\eta\frac{\partial\bar{w}^*}{\partial z} = -\frac{\rho a_2}{c_{44}s}\delta(z-h),$$

$$\tag{5.1.1}$$

$$i\kappa\xi\frac{\partial\bar{u}^*}{\partial z} + i\kappa\eta\frac{\partial\bar{v}^*}{\partial z} + \left(-\xi^2 - \eta^2 + \alpha\frac{\partial^2}{\partial z^2} - s^2\right)\bar{w}^* = -\frac{\rho a_3}{c_{44}s}\delta(z-h).$$

The (multiple) transform used here is given by, e.g.

$$\bar{u}^*(\xi, \eta, z, s) = \int_{-\infty}^{\infty} e^{-i\xi x}dx \int_{-\infty}^{\infty} e^{-i\eta y}dy \int_{0}^{\infty} e^{-s\tau}u(x, y, z, \tau)d\tau,$$

with inverse

$$u(x, y, z, \tau) = \frac{1}{(2\pi)^3 i} \int_{-\infty}^{\infty} e^{i\xi x}d\xi \int_{-\infty}^{\infty} e^{i\eta y}d\eta \int_{c} e^{s\tau}\bar{u}^*(\xi, \eta, z, s)ds. \tag{5.1.2}$$

The (transformed) boundary conditions require that

$$\bar{u}^*, \bar{v}^* \quad \text{and} \quad \bar{w}^* \to 0 \quad \text{as} \quad z \to \infty, \tag{5.1.3}$$

and on the stress free surface $z = 0$

$$\sigma_{13} = 0 \quad \text{or} \quad i\xi\bar{w}^* + \frac{\partial\bar{u}^*}{\partial z} = 0,$$

$$\sigma_{23} = 0 \quad \text{or} \quad i\eta\bar{w}^* + \frac{\partial\bar{v}^*}{\partial z} = 0, \tag{5.1.4}$$

$$\sigma_{33} = 0 \quad \text{or} \quad (\kappa-1)(i\xi\bar{u}^* + i\eta\bar{v}^*) + \alpha\frac{\partial\bar{w}^*}{\partial z} = 0.$$

The appearance of the Dirac deltas on the right side of equations (5.1.1) imply certain conditions of continuity and discontinuity at $z = h$. These conditions are

$$\bar{u}^*(\xi,\eta,h+0,s)-\bar{u}^*(\xi,\eta,h-0,s) = 0,$$

$$\bar{v}^*(\xi,\eta,h+0,s)-\bar{v}^*(\xi,\eta,h-0,s) = 0,$$

$$\bar{w}^*(\xi,\eta,h+0,s)-\bar{w}^*(\xi,\eta,h-0,s) = 0,$$

$$\frac{\partial}{\partial z}\bar{u}^*(\xi,\eta,h+0,s)-\frac{\partial}{\partial z}\bar{u}^*(\xi,\eta,h-0,s) = -\frac{pa_1}{c_{44}s},$$

$$\frac{\partial}{\partial z}\bar{v}^*(\xi,\eta,h+0,s)-\frac{\partial}{\partial z}\bar{v}^*(\xi,\eta,h-0,s) = -\frac{pa_2}{c_{44}s},$$

$$\frac{\partial}{\partial z}\bar{w}^*(\xi,\eta,h+0,s)-\frac{\partial}{\partial z}\bar{w}^*(\xi,\eta,h-0,s) = -\frac{pa_3}{\alpha c_{44}s}.$$

(5.1.5)

The solution to equations (5.1.1) satisfying the boundary conditions (5.1.3) and (5.1.4) will now be constructed by piecing together solutions to the corresponding homogeneous problem (i.e. with the right sides of equations (5.1.1) set to zero) which match the continuity/discontinuity conditions (5.1.5) at $z = h$. With this goal in mind, solutions to the homogeneous equations (5.1.1) will have the form

$$\bar{u}_H^* = Ue^{kz}, \bar{v}_H^* = Ve^{kz} \quad \text{and} \quad \bar{w}_H^* = We^{kz}, \tag{5.1.6}$$

where U, V, W and k are 'constants' (independent of z, but not necessarily of ξ, η or s) to be determined. Substitution of (5.1.6) into the homogeneous set (5.1.1) gives

$$(-\beta\xi^2 -\delta\eta^2 + k^2 - s^2)U-(\beta-\delta)\xi\eta V + i\kappa\xi kW = 0,$$

$$-(\beta-\delta)\xi\eta U + (-\delta\xi^2 - \beta\eta^2 + k^2 - s^2)V + i\kappa\eta kW = 0, \tag{5.1.7}$$

$$i\kappa\xi kU + i\kappa\eta kV + (-\xi^2 -\eta^2 + \alpha k^2 - s^2)W = 0.$$

Requiring the system determinant of equations (5.1.7) to vanish gives the sextic equation for k

$$[k^2 -\delta(\xi^2 +\eta^2)-s^2] \; [\alpha k^4 -\gamma k^2(\xi^2 +\eta^2)+ \beta(\xi^2 +\eta^2)^2 -(\alpha+1)k^2s^2$$

$$+ (\beta+1)(\xi^2 +\eta^2)s^2 + s^4] = 0. \tag{5.1.8}$$

Defining M by

$$M = \{[\gamma(\xi^2 +\eta^2)+(\alpha+1)s^2]^2 - 4\alpha[\beta(\xi^2 +\eta^2)^2$$

$$+ (\beta+1)(\xi^2 +\eta^2)s^2 + s^4]\}^{1/2} \tag{5.1.9}$$

allows the six k-roots of (5.1.8) to be written as

$$k_1 = \{\gamma(\xi^2 +\eta^2)+(\alpha+1)s^2 + M\}^{1/2}/\sqrt{2\alpha},$$

$$k_2 = -k_1,$$

$$k_3 = \{\gamma(\xi^2 +\eta^2)+(\alpha+1)s^2 - M\}^{1/2}/\sqrt{2\alpha},$$

$$k_4 = -k_3, \tag{5.1.10}$$

$$k_5 = [\delta(\xi^2 +\eta^2)+ s^2]^{1/2},$$

$$k_6 = -k_5.$$

Anticipating results to be established below in sections 5.2 and 5.3, the above k-roots will satisfy $\mathrm{Re}\, k_i > 0$ for $i = 1, 2, 3$ while $\mathrm{Re}\, k_i < 0$ for $i = 2, 4, 6$, these conditions holding everywhere on some suitable integration path.

Excepting two special cases (wherein u and v satisfy a fourth rather than a sixth order partial differential equation and for which the above k-quartic may be factorized) $\beta = \delta$, $\gamma = 1 + \alpha\beta$ and $\delta = 1$, $\gamma = \alpha + \beta$, the solution to the homogeneous equations, suggested by (5.1.6) is

$$\bar{u}_H^* = \sum_{n=1}^{6} U_n e^{k_n z}$$

$$\bar{v}_H^* = \sum_{n=1}^{6} V_n e^{k_n z}, \tag{5.1.11}$$

$$\bar{w}_H^* = \sum_{n=1}^{4} W_n e^{k_n z},$$

with $\xi U_n + \eta V_n = 0$ for $n = 5,6$. Of course the U, V and W's are not independent since they may be expressed as linear combinations of the six unknowns U_5, U_6, W_1, W_2, W_3 and W_4. Incidentally, the fact that $\bar{w}_H^*$ requires but four k-values rather than six, stems from the observation (see equation (1.7.3)) that the axial displacement component w satisfies a fourth order partial differential equation, whereas the transverse displacements u and v require sixth order operators to describe their motion.

The solution to the (nonhomogeneous) equations (5.1.1) may now be constructed from the homogeneous solutions (5.1.11) and will have the form

$$\bar{u}^* = \begin{cases} \displaystyle\sum_{n=1}^{6} U_n^{(1)} e^{k_n z} & \text{for} \quad 0 < z \leqslant h \\[4mm] \displaystyle\sum_{n=1}^{6} U_n^{(2)} e^{k_n z} & \text{for} \quad h \leqslant z < \infty, \end{cases} \tag{5.1.12}$$

with similar expressions for $\bar{v}^*$ and $\bar{w}^*$. At the present stage there are still 12 unknown quantities $U_5^{(i)}$, $U_6^{(i)}$, $W_1^{(i)}$, $W_2^{(i)}$, $W_3^{(i)}$, and $W_4^{(i)}$ for $i = 1, 2$. Twelve equations for these unknowns may be obtained from the three boundary conditions at infinity (5.1.3), the six continuity/discontinuity conditions at $z = h$ (5.1.5) plus the three surface boundary conditions (5.1.4).

In the next four sections, the surface motion at the epicenter will be discussed. For this reason, only the displacement coefficients appropriate to the region $0 < z \leqslant h$ will be recorded here.

Calling

$$A_n = \frac{(1 + \alpha\delta - \gamma)k_n^2 + (\beta - \delta)(\xi^2 + \eta^2) + (\beta - \delta)s^2}{\kappa k_n [k_n^2 - \delta(\xi^2 + \eta^2) - s^2]} \tag{5.1.13}$$

permits U_n and V_n to be expressed in terms of W_n by

$$U_n^{(1)} = -\, i\xi A_n W_n^{(1)} \quad \text{and} \quad V_n^{(1)} = -i\eta A_n W_n^{(1)} \quad \text{for} \quad n = 1, 2, 3, 4. \tag{5.1.14}$$

Then

$$U_5^{(1)} = U_6^{(1)} = \frac{\rho(-a_1\eta^2 + a_2\xi\eta)}{2c_{44}k_6 s(\xi^2 + \eta^2)} e^{-k_5 h} \tag{5.1.15}$$

$$V_5^{(1)} = V_6^{(1)} = \frac{\rho(a_1\xi\eta - a_2\xi^2)}{2c_{44}k_6 s(\xi^2 + \eta^2)} e^{-k_5 h}, \tag{5.1.16}$$

$$W_1^{(1)} = \frac{\rho}{2\alpha c_{44} s(k_4^2 - k_2^2)} \left[\frac{a_3 k_2 (p^2 - \alpha k_4^2)}{p^2} + i\kappa(a_1\xi + a_2\eta) \right] e^{-k_1 h}, \tag{5.1.17}$$

$$W_2^{(1)} = \frac{k_4 + k_2}{k_4 - k_2} \left[\frac{2(1-\kappa)(\xi^2 + \eta^2)p^2 - \{\gamma(\xi^2 + \eta^2) + \alpha s^2\} p^2 + \alpha s^2 k_2 k_4}{D} \right] W_1^{(1)}$$

$$+ \frac{2k_2}{\kappa(k_4 - k_2)} \left[\frac{\{(\kappa - 1)p^2 + \alpha k_4^2\}\{p^2 - \alpha k_4^2 - \kappa(\xi^2 + \eta^2)\}}{D} \right] W_3^{(1)}, \tag{5.1.18}$$

$$W_3^{(1)} = \frac{\rho}{2\alpha c_{44} s(k_4^2 - k_2^2)} \left[-\frac{a_3 k_4(p^2 - \alpha k_2^2)}{p^2} - i\kappa(a_1\xi + a_2\eta) \right] e^{-k_3 h}, \tag{5.1.19}$$

$$W_4^{(1)} = -\frac{2k_4}{\kappa(k_4 - k_2)} \left[\frac{\{(\kappa - 1)p^2 + \alpha k_2^2\}\{p^2 - \alpha k_2^2 - \kappa(\xi^2 + \eta^2)\}}{D} \right] W_1^{(1)}$$

$$- \frac{k_4 + k_2}{k_4 - k_2} \left[\frac{2(1-\kappa)(\xi^2 + \eta^2)p^2 - \{\gamma(\xi + \eta^2) + \alpha s^2\} p^2 + \alpha s^2 k_2 k_4}{D} \right] W_3^{(1)}. \tag{5.1.20}$$

In the above $p^2 = \xi^2 + \eta^2 + s^2$ and

$$D = 2(1-\kappa)(\xi^2 + \eta^2)p^2 - \{\gamma(\xi^2 + \eta^2) + \alpha s^2\} p^2 - \alpha s^2 k_2 k_4. \tag{5.1.21}$$

The quantity D here plays the role of the (three-dimensional) Rayleigh function for a transversely isotropic material. Equations (5.1.12)–(5.1.21) now complete the transform solution to the buried source problem for the upper region ($0 < z \leqslant h$) of the half-space.

From equation (5.1.12) the (transformed) axial displacement for $0 < z \leqslant h$ is given by

$$\bar{w}^* = W_1^{(1)} e^{k_1 z} + W_2^{(1)} e^{k_2 z} + W_3^{(1)} e^{k_3 z} + W_4^{(1)} e^{k_4 z}. \tag{5.1.22}$$

The first and third terms represent direct, free space, waves from the source (respectively, S and P waves in the isotropic case) whereas the second and fourth terms represent waves reflected from the surface (respectively, SS, PS and SP, PP waves in the isotropic case). The transform inversion of the various displacement components

161

$\bar{u}^*$, $\bar{v}^*$ and $\bar{w}^*$ presents a difficult problem due to the complicated structure of the k's. Asymptotic results could be obtained at a great distance from the source or near the wave front. Perhaps the only spatial location that will yield an explicit solution is the epicenter $x = y = z = 0$. The scope of the present investigation will now be narrowed so as to treat only the epicenter motion caused by a buried source having only a normal component (hence $a_1 = a_2 = 0$ and $u = v = 0$ at the epicenter). Thus the transverse displacement components are no longer needed. The above described situation will now be explored in some detail in the next few sections.

2. Transform Inversion at the Epicenter for Materials Satisfying Condition (1) of Table 11

At the surface $z = 0$, equation (5.1.22) becomes

$$\bar{w}^*(\xi, \eta, 0, s) = W_1^{(1)} + W_2^{(1)} + W_3^{(1)} + W_4^{(1)}, \tag{5.2.1}$$

or, using equations (5.1.18) and (5.1.20)

$$\bar{w}^* = \frac{2k_4}{(k_4 - k_2)D} \left[(1 - \kappa)(\xi^2 + \eta^2)p^2 - \{\gamma(\xi^2 + \eta^2) + \alpha s^2\} p^2 + \alpha k_2^2 p^2 \right.$$

$$\left. - \frac{1}{\kappa} \{(\kappa - 1)p^2 + \alpha k_2^2\} \{p^2 - \alpha k_2^2\} \right] W_1^{(1)}$$

$$- \frac{2k_2}{(k_4 - k_2)D} \left[(1 - \kappa)(\xi^2 + \eta^2)p^2 - \{\gamma(\xi^2 + \eta^2) + \alpha s^2\} p^2 + \alpha k_4^2 p^2 \right.$$

$$\left. - \frac{1}{\kappa} \{(\kappa - 1)p^2 + \alpha k_4^2\} \{p^2 - \alpha k_4^2\} \right] W_3^{(1)}. \tag{5.2.2}$$

As in section 4.2, and for essentially the same reason, it will be mathematically convenient to restrict the materials considered in this section to those which conform to conditions (4.2.16) (herein repeated)

(i)   if $\alpha > \beta > 1$ then $2\sqrt{\alpha\beta} \leqslant \gamma \leqslant (1 + \alpha\beta)$,

(ii)   if $\beta > \alpha > 1$ then $(\alpha + \beta) \leqslant \gamma \leqslant (1 + \alpha\beta)$, $\qquad$ (5.2.3)

(iii)   if $\alpha = \beta > 1$ then $2\alpha \leqslant \gamma \leqslant (1 + \alpha^2)$.

Conditions (5.2.3) are necessary and sufficient for the k-roots of the quartic factor of (5.1.8) to be real for $(\xi^2 + \eta^2) \geqslant 0$ and $s \geqslant 0$. The k-roots of the quadratic factor in (5.1.8) will be real provided $\delta > 0$, which is included in the strong ellipticity condition (1.5.24). The slowness sheets for the nine crystals (denoted by (1)) in Table 11 which satisfy (5.2.3) are locally convex where the symmetry axis (the p-axis) pierces the surface. This in turn implies that the corresponding wave front surface is also convex along the symmetry axis (the z-axis).

The incoming wave terms $W_1^{(1)}$ and $W_3^{(1)}$ will first be individually inverted, and then the combined result for the epicenter motion will be stated.

From equation (5.1.2)

$$w_1^{(1)}(0,0,0,\tau) = \frac{1}{(2\pi)^3 i} \int_{-\infty}^{\infty} d\xi \int_{-\infty}^{\infty} d\eta \int_c e^{s\tau} W_1^{(1)}(\xi,\eta,s)ds. \tag{5.2.4}$$

Let

$$\xi = sr\cos\phi, \qquad \eta = sr\sin\phi \tag{5.2.5}$$

then

$$\int_{-\infty}^{\infty} \int_{-\infty}^{\infty} d\xi d\eta \rightarrow \int_0^{\infty} s^2 r dr \int_0^{2\pi} d\phi. \tag{5.2.6}$$

Now from equations (5.1.10) it is clear that the k's depend on ξ and η only in the combination $(\xi^2 + \eta^2)$, hence the k's will not depend on ϕ (a consequence of transverse isotropy). Define $f_1(\omega)$ and $f_3(\omega)$ by

$$k_1 = sf_1(\omega) \quad \text{and} \quad k_3 = sf_3(\omega) \quad \text{where} \quad \omega = r^2. \tag{5.2.7}$$

(Note that f_1 and f_3 become identical with (respectively) ζ_1 and ζ_3 of (4.2.14) provided ω^2 in the latter expressions is replaced by ω.) Then using equation (5.1.17)

$$w_1^{(1)}(0,0,0,\tau) = \frac{1}{2(2\pi)^3 i} \int_0^{\infty} d\omega \int_0^{2\pi} d\phi \int_c \frac{\rho e^{s(\tau-f_1 h)}}{2\alpha c_{44}(f_3^2 - f_1^2)}$$

$$\times \left[\frac{-a_3 f_1(\omega + 1 - \alpha f_3^2)}{(\omega + 1)} \right] ds. \tag{5.2.8}$$

Performing the ϕ integration gives

$$w_1^{(1)}(0,0,0,\tau) = \frac{\rho a_3}{8\pi\alpha c_{44} h} I_1, \tag{5.2.9}$$

where

$$I_1 = h \int_0^{\infty} g_1(\omega) d\omega \frac{1}{2\pi i} \int_c e^{s(\tau-f_1 h)} ds \tag{5.2.10}$$

and

$$g_1(\omega) = \frac{[\omega + 1 - \alpha f_3^2(\omega)]f_1(\omega)}{(\omega + 1)[f_1^2(\omega) - f_3^2(\omega)]}. \tag{5.2.11}$$

Now, because of the imposed restrictions on α, β and γ, $f_1(\omega) > 0$ for all $\omega \geq 0$ so that the Laplace inversion is immediate (a consequence of the Cagniard technique, since the positive ω-axis is now the Cagniard path)

$$\frac{1}{2\pi i}\int_c e^{s(\tau -f_1 h)}\,ds = \delta[\tau -f_1 h],\qquad(5.2.12)$$

so that

$$I_1(\theta) = \int_0^\infty g_1(\omega)\delta[\theta -f_1(\omega)]\,d\omega,\qquad(5.2.13)$$

where

$$\theta = \tau/h.\qquad(5.2.14)$$

From (5.1.10) and (5.2.7), expressions for f_1 and f_3 may be found

$$\begin{aligned}
f_1(\omega) &= [\gamma\omega + (\alpha + 1)+ m]^{1/2}/\sqrt{2\alpha}\,,\\
f_3(\omega) &= [\gamma\omega + (\alpha + 1)- m]^{1/2}/\sqrt{2\alpha}\,,
\end{aligned}\qquad(5.2.15)$$

where

$$m(\omega) = \{[\gamma\omega + (\alpha + 1)]^2 - 4\alpha[\beta\omega^2 + (\beta + 1)\omega + 1]\}^{1/2}.\qquad(5.2.16)$$

Now $f_1(0)=1$ and $f_1(\omega)$ is an increasing function with $f'_1(\omega)\neq 0$ on $0\leqslant\omega<\infty$. Consequently $\theta -f_1(\omega)$ has but one ω zero (say ω_1) in the integration range. Hence

$$I_1(\theta) = \frac{g_1(\omega_1)}{f'_1(\omega_1)}H(\theta - 1),\qquad(5.2.17)$$

where H is the Heaviside unit step function, and ω_1 and ω_3 are defined by

$$\begin{aligned}
\omega_1 &= [\gamma\theta^2 -(\beta + 1)-S(\theta)]/2\beta,\\
\omega_3 &= [\gamma\theta^2 -(\beta + 1)+S(\theta)]/2\beta,
\end{aligned}\qquad(5.2.18)$$

where

$$S(\theta) = \{[\gamma\theta^2 -(\beta + 1)]^2 - 4\beta[\alpha\theta^4 -(\alpha + 1)\theta^2 + 1]\}^{1/2}.\qquad(5.2.19)$$

The explicit expression for I_1 is

$$I_1(\theta) = 2\alpha\left[\frac{1}{2}+\frac{(2-\gamma)\theta^2 +(\beta - 1)}{2S(\theta)}\right]H(\theta - 1).\qquad(5.2.20)$$

The $W_3^{(1)}(\xi, \eta, s)$ term can be similarly inverted. Call

$$W_3^{(1)}(0,0,0,\tau) = \frac{\rho a_3}{8\pi\alpha c_{44}h}\,I_3\qquad(5.2.21)$$

where

$$I_3 = h\int_0^\infty g_3(\omega)d\omega\,\frac{1}{2\pi i}\int_c e^{s(\tau -f_3 h)}\,ds,\qquad(5.2.22)$$

and

$$g_3(\omega) = \frac{[\omega + 1 - \alpha f_1^2(\omega)] f_3(\omega)}{(\omega + 1) [f_3^2(\omega) - f_1^2(\omega)]},$$ (5.2.23)

then

$$I_3(\theta) = 2\alpha \left[\frac{1}{2} - \frac{(2 - \gamma)\theta^2 + (\beta - 1)}{2S(\theta)}\right] H(\theta - \alpha^{-1/2}).$$ (5.2.24)

The combined term $w_1^{(1)} (0, 0, 0, \tau) + w_3^{(1)} (0, 0, 0, \tau)$ represents the axial displacement for an infinite transversely isotropic elastic solid (with h corresponding to the axial distance from the source). The problem of axial motion of an unbounded transversely isotropic elastic solid has been treated earlier in section 3.11. With but minor notational changes, the expressions for $I_1(\theta)$ and $I_3(\theta)$ are in agreement with the results of section 3.11 and in fact the inversion technique of the present section gives an alternate approach to the free space axial motion problem.

The complete epicenter normal displacement can now be easily inverted and expressed in terms of I_1 and I_3. Toward this goal define

$$\frac{4\pi c_{44} h}{\rho a_3} w(0, 0, 0, \tau) = w_0(\theta),$$ (5.2.25)

where $\bar{w}^*$ is given by equation (5.2.2). Then

$$w_0(\theta) = \frac{I_1(\theta)F(\omega_1)}{2\alpha} + \frac{I_3(\theta)F(\omega_3)}{2\alpha}$$ (5.2.26)

where

$$A(\omega) = \frac{2[(\beta\omega + 1)(\omega + 1)]^{1/2}}{[(\beta\omega + 1)(\omega + 1)]^{1/2} - \theta^2\sqrt{\alpha}},$$ (5.2.27)

and

$$F(\omega) = \frac{A(\omega)}{B(\omega)} [(1 - \kappa)\omega(\omega + 1) - (\gamma\omega + \alpha)(\omega + 1) + \alpha(\omega + 1)\theta^2$$

$$- \frac{1}{\kappa} \{(\kappa - 1)(\omega + 1) + \alpha\theta^2\} \{\omega + 1 - \alpha\theta^2\}],$$ (5.2.28)

with

$$B(\omega) = [2(1 - \kappa)\omega(\omega + 1) - (\gamma\omega + \alpha)(\omega + 1) - \{\alpha(\beta\omega + 1)(\omega + 1)\}^{1/2}].$$ (5.2.29)

Some features of the vertical epicenter displacement $w_0(\theta)$, will be discussed below in section 5.5.

3. Discussion of the ω-Plane Branch Points and the Cagniard Path When the Real ω-Axis is Not Free of Branch Points

For the four remaining hexagonal crystals listed in Table 11,

$$(\gamma^2 - 4\alpha\beta) < 0 \quad \text{and} \quad \gamma < (\beta + \alpha). \tag{5.3.1}$$

These inequalities hold for apatite, beryllium, cadmium and zinc.

The expression (5.2.16) for $m(\omega)$ may be written as

$$m(\omega) = [(\gamma^2 - 4\alpha\beta)(\omega - \omega_-)(\omega - \omega_+)]^{1/2}, \tag{5.3.2}$$

where

$$\omega_\pm = \frac{-[\gamma(\alpha + 1) - 2\alpha(\beta + 1)] \mp 2[\alpha(\alpha + \beta - \gamma)(1 + \alpha\beta - \gamma)]^{1/2}}{(\gamma^2 - 4\alpha\beta)}, \tag{5.3.3}$$

as introduced earlier in equation (4.4.3). By virtue of (1.5.48) and (5.3.1), $m(\omega)$ will have branch points on the real ω-axis at $\omega = \omega_+$ and $\omega = \omega_-$. Since $\omega_+\omega_- = (\alpha - 1)^2/(\gamma^2 - 4\alpha\beta) < 0$, it follows that $\omega_+ > 0 > \omega_-$. Branch cuts for $m(\omega)$ are taken along the real ω-axis from ω_+ to $-\infty$ and ω_- to $-\infty$ with $-\pi < \arg(\omega - \omega_\pm) < \pi$ and $\arg(\gamma^2 - 4\alpha\beta) = -\pi$. Thus, effectively, the branch cut for $m(\omega)$ extends from ω_- to ω_+ along the real ω-axis and $m(0 + i0) = \alpha - 1$.

The functions $f_1(\omega)$ and $f_3(\omega)$ defined in (5.2.15) introduce further branch points at $\omega = -1$ and $\omega = -1/\beta$. From a consideration of $\gamma\omega + \alpha + 1 \pm m(\omega)$ evaluated at $\omega = -1$ and $\omega = -1/\beta$, together with the fact that $\omega_- < -1$ for the four crystals under investigation, it is easily established that for apatite and zinc f_1 has branch points at $\omega = \omega_-, \omega_+$ while f_3 has branch points at $\omega = \omega_-, \omega_+, -1, -1/\beta$. On the other hand, for beryllium and cadmium f_1 has branch points at $\omega = \omega_-, \omega_+, -1$ while f_3 has branch points at $\omega = \omega_-, \omega_+$ and $-1/\beta$. Taking apatite as an example

$$f_1(\omega) = |f_1(\omega)|^{1/2} e^{i\phi/2} \quad \text{with} \quad -\pi < \phi < \pi, \tag{5.3.4}$$

so that $\operatorname{Re} f_1(\omega) > 0$ on $0 \leqslant \omega < \infty$. Then for f_3 define a function G by

$$G(\omega) = \frac{[\gamma\omega + \alpha + 1 - m(\omega)]}{\left[2\alpha(\omega + 1)\left(\omega + \dfrac{1}{\beta}\right)\right]} \quad \text{for} \quad \omega \neq -1, -1/\beta, \tag{5.3.5}$$

with $G(-1)$ and $G(-1/\beta)$ defined by the limit of G at these removable singularities. Also set $\phi = \arg G$, then

$$f_3(\omega) = |G(\omega)|^{1/2} e^{i\phi/2} [(\omega + 1)(\omega + 1/\beta)]^{1/2}, \tag{5.3.6}$$

where

$$\omega + 1 = |\omega + 1| e^{i\phi_1}, \quad -\pi < \phi_1 < \pi, \tag{5.3.7}$$

and

$$\omega + 1/\beta = |\omega + 1/\beta| e^{i\phi_2}, \quad -\pi < \phi_2 < \pi. \tag{5.3.8}$$

With the f_3 branch cuts so taken, $\operatorname{Re} f_3(\omega) > 0$ on $0 \leqslant \omega < \infty$ also. Branch cuts

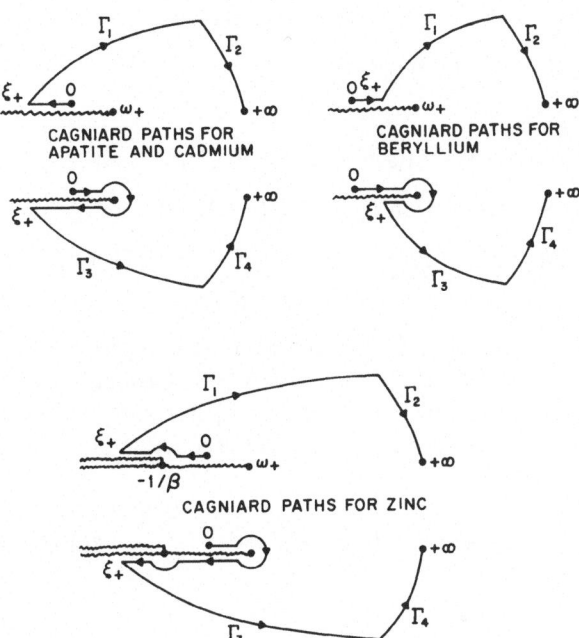

Figure 70. Complex ω-Plane Integration Paths

associated with beryllium and cadmium can be similarly defined, with the result that the real part of f_1 and f_3 remain positive along the original, ω-plane, integration path as shown at the top of figure 70. The requirement that $\mathrm{Re}\, f_1 > 0$ and $\mathrm{Re}\, f_3 > 0$ on the ω-plane path is implicit in the integral transform solution given in section 5.1. Incidentally, this requirement can be met by choosing the branch cuts in other ways, but (anticipating a result to be given below) only cuts along the real axis can conveniently serve as portions of the Cagniard path.

The Cagniard integration path for the ω-integral in I_1 of (5.2.10) is found by setting

$$f_1(\omega) = T, \qquad T = \text{real}, \tag{5.3.9}$$

where ω is now a complex variable given by

$$\omega = \xi + i\eta. \tag{5.3.10}$$

Using (5.2.15) and (5.3.10) in (5.3.9), squaring twice and equating the real and imaginary part of the equation gives

$$\gamma T^2 \xi + (\alpha + 1)T^2 - \alpha T^4 = \beta(\xi^2 - \eta^2) + (\beta + 1)\xi + 1, \tag{5.3.11}$$

$$\gamma T^2 \eta = 2\beta \xi \eta + (\beta + 1)\eta. \tag{5.3.12}$$

167

From (5.3.12) one possibility is $\eta = 0$, for which (5.3.11) imposes the additional condition

$$-1/\beta \leqslant \xi \leqslant \omega_+, \tag{5.3.13}$$

if $-1/\beta$ is a branch point, as it may be for the Cagniard path associated with f_3. On the other hand, if $\eta \neq 0$, then (5.3.11) and (5.3.12) imply

$$\xi(T) = [\gamma T^2 - (\beta + 1)]/2\beta, \tag{5.3.14}$$

and

$$\eta(T) = \{(4\alpha\beta - \gamma^2)^{1/2}[(T^2 - T_+^2)(T^2 - T_-^2)]^{1/2}\}/2\beta, \tag{5.3.15}$$

where $T_+ \leqslant T < \infty$. The notation $T_\pm$ used in (5.3.15) is

$$T_\pm^2 = \frac{-[\gamma(\beta + 1) - 2\beta(\alpha + 1)] \pm [4\beta(\alpha + \beta - \gamma)(1 + \alpha\beta - \gamma)]^{1/2}}{(4\alpha\beta - \gamma^2)}. \tag{5.3.16}$$

Equations (5.3.14) and (5.3.15) give the parametric representation of the Cagniard integration path which is off the real ω-axis and is associated with the exponential term involving $f_1(\omega)$. This path is denoted by Γ_1 in Figure 70. Elimination of the parameter T between (5.3.14) and (5.3.15) gives the equation of a hyperbola,

$$\left[\xi + \frac{2\alpha(\beta + 1) - \gamma(\alpha + 1)}{4\alpha\beta - \gamma^2}\right]^2 - \frac{\gamma^2\eta^2}{4\alpha\beta - \gamma^2} = \frac{\gamma^2(\alpha + \beta - \gamma)(1 + \alpha\beta - \gamma)}{\beta(4\alpha\beta - \gamma^2)^2} \tag{5.3.17}$$

This hyperbola intersects the ξ-axis at the two points $\xi = \xi_\pm$ where

$$\xi_\pm = \frac{\gamma(\alpha + 1) - 2\alpha(\beta + 1) \pm \gamma[(\alpha + \beta - \gamma)(1 + \alpha\beta - \gamma)/\beta]^{1/2}}{(4\alpha\beta - \gamma^2)}. \tag{5.3.18}$$

Note that $\xi_+ = \xi(T_+)$. The contour Γ_1 is thus the upper right portion of the hyperbola (5.3.17). Similarly the Cagniard path for integral I_3, associated with the exponential term $f_3(\omega)$, may involve a portion of the ξ-axis together with the contour

$$\Gamma_3: \quad \omega = \xi(T) - i\eta(T), \tag{5.3.19}$$

which is the lower right portion of the hyperbola (5.3.17). The fact that Γ_1 is the upper part and Γ_3 the lower part of the right branch of the hyperbola, and not vice versa, may be established from a consideration of the asymptotic form of (5.3.14) and (5.3.15) as $T \to \infty$ and $f_1(\omega)$ as $\omega \to \infty + i\infty$.

The Cagniard integration paths which replace the original integration path are shown in Figure 70. Each contour must be completed by a (non-Cagniard path) arc of a circle at infinity. These arcs are denoted by Γ_2 and Γ_4. Note that the contour for zinc is somewhat special in that $-1/\beta > \xi_+$ (which will have important consequences, to be discussed below in sections 5.5 and 5.6). Now $-1/\beta$ is not a branch point of f_1 for zinc, so that the upper contour shown in Figure 70 is all right. However, the lower contour for zinc involves a portion of the ξ-axis where $\xi_+ < \xi < -1/\beta$, which apparently is not a Cagniard path for $f_3(\omega)$, by (5.3.13), since $-1/\beta$ is a branch point for

f_3. This trouble is only apparent, since $m(\omega)$ changes sign across the cut and hence $f_3(\omega)$ becomes $f_1(\omega)$ there.

There is an additional pole singularity introduced into the complex ω-plane, caused by the zero of D in (5.1.21). When D is expressed in terms of ω, it is found that D has a simple root on the negative ξ-axis to the left of $\xi = -1$. This is the Rayleigh pole $\omega = \xi_R$ (hence $\xi_R = (i\omega_R)^2$ in the notation of Table 13), associated with surface waves. This singularity does not have any direct influence on the epicenter motion caused by a buried source. Table 14 summarizes some numerical data for this section.

Table 14. Numerical data for Cagniard integration paths.

Crystal	ω_+	ω_-	$-1/\beta$	ξ_+	ξ_R
Apatite	0.075383	− 1.034910	− 0.396825	− 0.198086	− 1.359327
Beryllium	0.478478	− 1.072811	− 0.555556	+ 0.418089	− 1.342879
Cadmium	0.103825	− 1.568456	− 0.168067	− 0.012288	− 1.150147
Zinc	0.015883	− 1.001393	− 0.239808	− 0.254209	− 1.296415

4. Transform Inversion at the Epicenter for Materials Satisfying Conditions (2) and (3) of Table 11.

The original integration path for I_1 (see (5.2.10)) can be deformed into the top contour appropriate to apatite as shown in Figure 70. Thus

$$I_1 = I_{13} + I_{11} + I_{12},\tag{5.4.1}$$

where

$$I_{13} = h \int_0^{\xi_+} g_1(\omega)d\omega \frac{1}{2\pi i} \int_c e^{s(\tau - hf_1)} ds,\tag{5.4.2}$$

$$I_{11} = h \int_{\Gamma_1} g_1(\omega)d\omega \frac{1}{2\pi i} \int_c e^{s(\tau - hf_1)} ds\tag{5.4.3}$$

and

$$I_{12} = h \int_{\Gamma_2} g_1(\omega)d\omega \frac{1}{2\pi i} \int_c e^{s(\tau - hf_1)} ds.\tag{5.4.4}$$

The integral I_{12} tends to zero as $|\omega| \to \infty$ over the circular arc Γ_2, so that

$$I_{12} = 0.\tag{5.4.5}$$

On Γ_1,

$$f_1(\omega) = T \quad \text{and} \quad T_+ \leqslant T < \infty,\tag{5.4.6}$$

so that

$$I_{11} = h \int_{T_+}^{\infty} \frac{2\alpha T^2 \left[\omega + 1 + \alpha T^2 - \gamma\omega - \alpha - 1\right]}{(\omega + 1) \left[\gamma T^2 - (\beta + 1 + 2\beta\omega)\right]} dT \frac{1}{2\pi i} \int_c e^{s(\tau - hT)} ds,$$

$$(5.4.7)$$

where $\omega = \xi(T) + i\eta(T)$. The double integration is now easily performed and results in

$$I_{11} = 2\alpha \left[\frac{1}{2} + i \frac{(2-\gamma)\theta^2 + (\beta-1)}{2Q(\theta)}\right] H(\theta - T_+),$$

$$(5.4.8)$$

where

$$Q(\theta) = \{4\beta(\alpha\theta^2 - 1)(\theta^2 - 1) - [\gamma\theta^2 - (\beta+1)]^2\}^{1/2},$$

$$(5.4.9)$$

and $\theta = \tau/h$ as previously defined in (5.2.14). Note that $Q(T_+) = 0$. The wave arrival time $\theta = T_+$, or $\tau = hT_+$, corresponds to the arrival at the epicenter of the conical point on the free space wave front from the source at depth h.

After the Laplace inversion, the integral I_{13} of (5.4.2) may be written

$$I_{13} = -\int_{\xi_+}^{0} g_1(\omega)\delta\left[\theta - f_1(\omega)\right] d\omega = -\frac{g_1(\omega_3)}{|f_1'(\omega_3)|} \quad \text{for} \quad 1 \leqslant \theta < T_+.$$

$$(5.4.10)$$

If the expression for ω_3 given in (5.2.18) is inserted in (5.4.10) the result is

$$I_{13} = 2\alpha \left[\frac{1}{2} - \frac{(2-\gamma)\theta^2 + (\beta-1)}{2S(\theta)}\right] \quad \text{for} \quad 1 \leqslant \theta \leqslant T_+.$$

$$(5.4.11)$$

Note that $S(T_+) = 0$.

The deformed integration path for I_3 of (5.2.22) is shown as the lower contour for apatite in Figure 70. Thus

$$I_3 = I_{31} + \oint_{\omega_+} + I_{32} + I_{331} + I_{33} + I_{34},$$

$$(5.4.12)$$

where

$$I_{31} = h \int_0^{\omega_+} g_3(\omega)d\omega \frac{1}{2\pi i} \int_c e^{s(\tau - hf_3)} ds$$

$$= \int_0^{\omega_+} g_3(\omega)\delta\left[\theta - f_3(\omega)\right] d\omega,$$

$$(5.4.13)$$

$$\oint_{\omega_+} \to 0 \quad \text{as} \quad |\omega| \to 0,$$

$$(5.4.14)$$

$$I_{32} = h \int_{\omega_+}^0 g_3(\omega)d\omega \frac{1}{2\pi i} \int_c e^{s(\tau - hf_3)}ds, \quad \text{where} \quad \omega = \xi - i0. \tag{5.4.15}$$

But, as $m \to m$ below the branch cut, $f_1(\omega)$ and $f_3(\omega)$ are interchanged as are $g_1(\omega)$ and $g_3(\omega)$ so that

$$I_{32} = -\int_0^{\omega_+} g_1(\omega)\delta[\theta - f_1(\omega)]\,d\omega, \tag{5.4.16}$$

$$I_{331} = h \int_0^{\xi_+} g_1(\omega)\delta[\theta - f_1(\omega)]\,d\omega = I_{13}, \tag{5.4.17}$$

where I_{13} is given in (5.4.11),

$$I_{33} = h \int_{\Gamma_3} g_3(\omega)d\omega \frac{1}{2\pi i} \int_c e^{s(\tau - hf_3)}ds \tag{5.4.18}$$

and

$$I_{34} = h \int_{\Gamma_4} g_3(\omega)d\omega \frac{1}{2\pi i} \int_c e^{s(\tau - hf_3)}ds. \tag{5.4.19}$$

Again, $I_{34} = 0$, since the integrand vanishes as $|\omega| \to \infty$ on the circular arc Γ_4. The above integral associated with I_3 may be evaluated in a manner similar to those for I_1. The results are

$$I_{31} + I_{32} = 2\alpha \left[\frac{1}{2} - \frac{(2-\gamma)\theta^2 + (\beta-1)}{2S(\theta)} \right] \quad \text{for} \quad \alpha^{-1/2} < \theta \leqslant 1 \tag{5.4.20}$$

and

$$I_{33} = \text{complex conjugate of } I_{11}. \tag{5.4.21}$$

Define

$$w_1(\theta) = \frac{4\pi c_{44} h}{\rho a_3} [w_1^{(1)}(0,0,0,\tau) + w_3^{(1)}(0,0,0,\tau)], \tag{5.4.22}$$

then for

$$
\begin{aligned}
T_+ \leqslant \theta < \infty, & \qquad w_1(\theta) = 1, \\
1 \leqslant \theta < T_+, & \qquad w_1(\theta) = 2g(\theta), \\
\alpha^{-1/2} < \theta \leqslant 1, & \qquad w_1(\theta) = g(\theta), \\
0 \leqslant \theta \leqslant \alpha^{-1/2}, & \qquad w_1(\theta) = 0,
\end{aligned}
\tag{5.4.23}
$$

where

$$g(\theta) = \left[\frac{1}{2} - \frac{(2-\gamma)\theta^2 + (\beta-1)}{2S(\theta)} \right]. \tag{5.4.24}$$

The expressions in (5.4.23) represent the free space axial motion at a distance h from the source. This result for apatite, is also valid for cadmium and zinc and has been previously obtained by a different method in section 3.11.

Return now to the main problem of evaluating (5.2.2). Using the definition of $w_0(\theta)$ given in (5.2.25) and employing previous integration variable changes of this section gives,

$$w_0(\theta) = \frac{h}{2a} \int_0^\infty F_1(\omega)g_1(\omega)d\omega \, \frac{1}{2\pi i} \int_c e^{s(\tau - hf_1)} ds$$

$$+ \frac{h}{2a} \int_0^\infty F_3(\omega)g_3(\omega)d\omega \, \frac{1}{2\pi i} \int_c e^{s(\tau - hf_3)} ds, \tag{5.4.25}$$

where

$$F_1(\omega) = \frac{2f_3 V_1}{(f_3 - f_1)d_0}, \qquad F_3(\omega) = \frac{2f_1 V_3}{(f_1 - f_3)d_0}, \tag{5.4.26}$$

$$d_0(\omega) = [2(1-\kappa)\omega(\omega+1) - (\gamma\omega + \alpha)(\omega+1) - \{\alpha(\beta\omega+1)(\omega+1)\}^{1/2}] \tag{5.4.27}$$

and

$$V_j(\omega) = [(1-\kappa)\omega(\omega+1) - (\gamma\omega + \alpha)(\omega+1) + \alpha(\omega+1)f_j^2$$

$$- \frac{1}{\kappa}\{(\kappa-1)(\omega+1) + \alpha f_j^2\}\{\omega + 1 - \alpha f_j^2\}] \qquad j = 1, 2 \tag{5.4.28}$$

The integration now proceeds as above for w_1.

Summary for Apatite and Cadmium Epicenter Normal Displacement.

$$T_+ \leqslant \theta < \infty, \qquad w_0(\theta) = 2\mathrm{Re}\left\{F(\omega)\left[\frac{1}{2} + i\frac{(2-\gamma)\theta^2 + (\beta-1)}{2Q(\theta)}\right]\right\},$$

where $\omega = \xi(\theta) + i\eta(\theta)$ see (5.3.14) and (5.3.15)

$$1 \leqslant \theta < T_+, \qquad w_0(\theta) = 2F(\omega_3)g(\theta),$$

$$\alpha^{-1/2} < \theta \leqslant 1, \qquad w_0(\theta) = F(\omega_3)g(\theta), \tag{5.4.29}$$

$$0 \leqslant \theta \leqslant \alpha^{-1/2}, \qquad w_0(\theta) = 0.$$

In the above, ω_3 has been defined in (5.2.18) and

$$F(\omega) = \frac{2fV}{(f-\theta)d} \quad \text{with} \quad f(\omega) = [(\gamma\omega + \alpha + 1 - \alpha\theta^2)/\alpha]^{1/2}, \tag{5.4.30}$$

$$V(\omega) = [(1-\kappa)\omega(\omega+1) - (\gamma\omega + \alpha)(\omega+1) + \alpha\theta^2(\omega+1)$$

$$- \frac{1}{\kappa}\{(\kappa-1)(\omega+1) + \alpha\theta^2\}\{\omega + 1 - \alpha\theta^2\}], \tag{5.4.31}$$

and

$$d(\omega) = [2(1 - \kappa)\omega(\omega + 1) - (\gamma\omega + \alpha)(\omega + 1) - \alpha\theta f]. \qquad (5.4.32)$$

This concludes the analysis for apatite and cadmium.

In Table 14 and also in Figure 70, it is seen that ξ_+ is positive for beryllium. This has a profound effect on the solution. For example, the free space solution will contain a singularity at $\theta = T_+$ if $\xi_+ < 0$, whereas if $\xi_+ > 0$ no singularity is present. A glance at a quadrant of the slowness curve for beryllium and zinc, shown in Figure 71, will help to explain this distinction. The equation of the slowness curve, from (2.2.3) is

$$F(p, q) = \alpha p^4 + \gamma p^2 q^2 + \beta q^4 - (\alpha + 1)p^2 - (\beta + 1)q^2 + 1 = 0, \qquad (5.4.33)$$

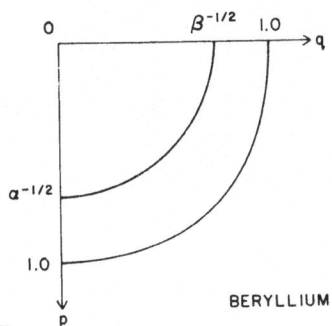

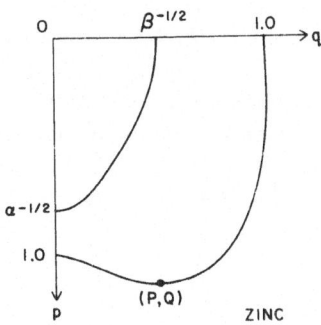

Figure 71. Slowness Curves for Beryllium and Zinc

where the p-axis is the symmetry axis. If (P, Q) denote the coordinates of the off-axis point on the slowness curve where the slope is zero, then this point is found at the intersection of the quartic (5.4.33) and the ellipse

$$2\beta q^2 + \gamma p^2 - (\beta + 1) = 0. \qquad (5.4.34)$$

By calculating the coordinates of this intersection point and comparing with (5.3.16) and (5.3.18) it is seen that

$$P = T_+ \quad \text{and} \quad Q^2 = -\xi_+. \qquad (5.4.35)$$

But since ξ_+ is positive for beryllium, the slowness curve has no off-axis horizontal tangent. Thus in the present context $\xi_+ = 0$, which corresponds to $\gamma = \beta + 1$, gives the critical value between singular and nonsingular solutions for the axial displacement.

With minor changes the above analysis can be extended to beryllium. The free space solution is identical with that recorded in section 5.2 and 3.11 and need not be listed here.

Epicenter Normal Displacement for Beryllium.

$$T_+ \leqslant \theta < \infty, \qquad w_0(\theta) = 2\mathrm{Re}\left\{F(\omega)\left[\frac{1}{2} + i\frac{(2-\gamma)\theta^2 + (\beta-1)}{2Q(\theta)}\right]\right\}$$

$$\text{where } \omega = \xi(\theta) + i\eta(\theta),$$

$$1 \leqslant \theta \leqslant T_+, \qquad w_0(\theta) = F(\omega_1)k(\theta) + F(\omega_3)g(\theta), \tag{5.4.36}$$

$$\alpha^{-1/2} < \theta \leqslant 1, \qquad w_0(\theta) = F(\omega_3)g(\theta),$$

$$0 \leqslant \theta \leqslant \alpha^{-1/2}, \quad w_0(\theta) = 0,$$

where ω_1 is given in (5.2.18) and

$$k(\theta) = \left[\frac{1}{2} + \frac{(2-\gamma)\theta^2 + (\beta-1)}{2S(\theta)}\right]. \tag{5.4.37}$$

The beryllium solution for the interval $T_+ \leqslant \theta < \infty$ is clearly just the analytic extension of the solution for $1 \leqslant \theta \leqslant T_+$ formed by $i\bar{\eta} \rightarrow -S(\theta)/2\beta$ and $iQ \rightarrow -S(\theta)$. With this remark, it is seen that the beryllium solution conforms to that found in section 5.2.

The integration contour for zinc contains the additional feature that $-1/\beta > \xi_+$. This causes no modification of the free space solution, which is still given by (5.4.23), but does alter the epicenter solution form in a most interesting way to be explained below in sections 5.5 and 5.6.

Epicenter Normal Displacement for Zinc.

$$T_+ < \theta < \infty, \qquad w_0(\theta) = 2\mathrm{Re}\left\{F(\omega)\left[\frac{1}{2} + i\frac{(2-\gamma)\theta^2 + (\beta-1)}{2Q(\theta)}\right]\right\}$$

$$\text{where } \omega = \xi(\theta) + i\eta(\theta),$$

$$1 \leqslant \theta < T_+, \qquad w_0(\theta) = 2\mathrm{Re}\left\{F(\omega_3)g(\theta)\right\}, \tag{5.4.38}$$

$$\alpha^{-1/2} < \theta \leqslant 1, \qquad w_0(\theta) = F(\omega_3)g(\theta),$$

$$0 \leqslant \theta \leqslant \alpha^{-1/2}, \quad w_0(\theta) = 0.$$

5. Discussion of the Epicenter Vertical Displacement for Some Hexagonal Crystals

Although the expression (5.2.26) for $w_0(\theta)$ (appropriate to the nine crystals marked (1) in Table 11) is explicit, it is not easy to diagnose the shape of the curve w_0 vs. θ. Numerical results help in this respect. The main numerical features of $w_0(\theta)$ are summarized in Table 15 for nine hexagonal crystals.

Table 15. Numerical features of $w_0(\theta)$ for the (1) crystals of Table 11.

Crystal	$\alpha^{-1/2}$	$w_0(\alpha^{-1/2} + 0)$	$w_0(1)$	$-J$	θ_M	$w_0(\theta_M)$	$w_0(\infty)$
Beryl	0.525588	0.783469	2.322487	0.438018	2.70	2.417610	2.410323
Cobalt	0.459315	0.800702	2.295953	0.286945	2.35	2.355557	2.347279
Ice	0.467780	0.678179	2.269672	0.450177	2.27	2.341740	2.330572
Hafnium	0.531494	0.693515	2.321753	1.825865	2.29	2.452143	2.436742
Magnesium	0.517088	0.681473	2.303496	1.139825	2.31	2.415590	2.401856
Rhenium	0.486792	0.766521	2.310685	0.498601	2.35	2.389547	2.379488
Titanium	0.507673	0.637620	2.283518	1.831316	2.18	2.406124	2.388355
Thallium	0.370879	0.401957	2.129994	0.784798	1.95	2.212321	2.192591
Yttrium	0.562544	0.790919	2.371430	1.247848	2.63	2.509984	2,500366

The vertical displacement at the epicenter is zero prior to the arrival of the first disturbance at $\theta = \alpha^{-1/2}$ $(\tau = h\alpha^{-1/2})$ whereupon it suddenly jumps to the value $w_0(\alpha^{-1/2} + 0)$ where

$$w_0(\alpha^{-1/2} + 0) = \frac{2(\alpha - 1)}{\alpha(\beta + 1) - \gamma} . \tag{5.5.1}$$

The vertical displacement now steadily increases until the second disturbance arrives at $\theta = 1 (\tau = h)$ where w_0 has the value

$$w_0(1) = \frac{2(1 - \kappa - \gamma)\,[(\gamma - \beta)\,(\gamma - 1) + \{\alpha\beta(\gamma - \beta)\,(\gamma - 1)\}^{1/2}]}{(\gamma - 1)\,\{(2 - 2\kappa - \gamma)\,(\gamma - \beta - 1) - \alpha\beta\} - \beta\{\alpha\beta(\gamma - \beta)\,(\gamma - 1)\}^{1/2}} . \tag{5.5.2}$$

The function $w_0(\theta)$ is continuous at $\theta = 1$, but suffers a jump in slope given by

$$J = w_0'(1 + 0) - w_0'(1 - 0), \tag{5.5.3}$$

where

$$J = -\frac{4\kappa(\kappa - 1 + \alpha)}{\sqrt{\alpha}\,(\beta + 1 - \gamma)^2} \tag{5.5.4}$$

The vertical displacement continues to increase until $\theta = \theta_M$ where w_0 takes on its maximum value. After this time w_0 slowly relaxes to its static value

$$w_0(\infty) = \frac{2(\kappa - 1 + \gamma + \sqrt{\alpha\beta})}{2\kappa + \gamma - 2} . \tag{5.5.5}$$

For the crystals, whose properties are tabulated in Table 15, the static and maximum

175

values of the normal displacement differ only by about 1%, hence $w_0(\theta)$ is a fair estimate of the maximum vertical rise of the surface at the epicenter.

Figure 72 shows a graph of the epicenter motion for magnesium. The features of this graph, as described above, are common to all the crystals listed in Table 15.

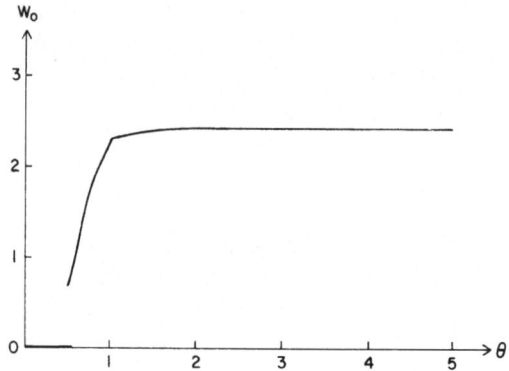

Figure 72.   Vertical Displacement at the Epicenter for Magnesium.

It remains now to discuss the epicenter displacement features for the materials designated by a (2) or (3) in Table 11. Equations (5.4.29) for apatite and cadmium, (5.4.36) for beryllium and (5.4.38) for zinc are more or less explicit expressions for $w_0(\theta)$ and can be made more explicit by taking the appropriate real part. This operation would, perhaps, double the size of the formulae without greatly increasing their intelligibility. Fortunately complex algebra can now be routinely handled by Fortran, and this was used on Adelphi's Burroughs computer to obtain data for the four graphs shown in Figure 73.

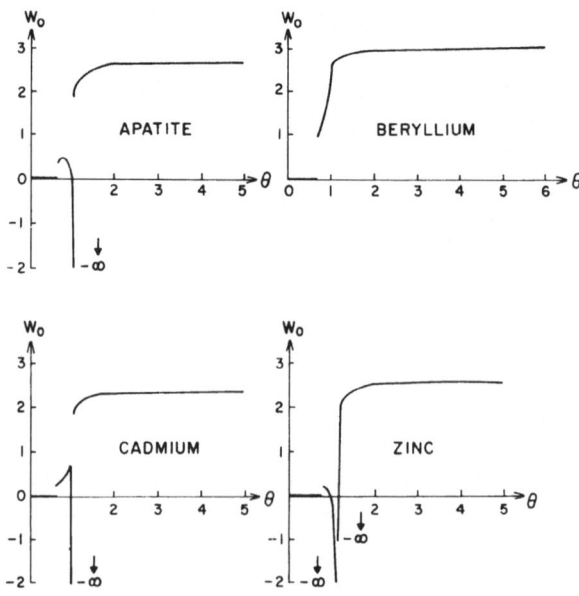

Figure 73.   Vertical Displacement at the Epicenter for Apatite, Beryllium Cadmium and Zinc.

5. Discussion of the Epicenter Vertical Displacement

The vertical displacement at the epicenter is zero prior to the arrival of the first disturbance at $\theta = \alpha^{-1/2}$ whereupon it suddenly jumps to the value $w_0(\alpha^{-1/2} + 0)$ as given in (5.5.1) The graphs for apatite and cadmium continue to rise to a local maximum value before a sharp decline to zero at the arrival of the second wave front when $\theta = 1$. The graph for zinc decreases monotonically to zero, while the graph for beryllium rises monotonically to the value $w_0(1)$ given by equation (5.5.2). Again the function $w_0(\theta)$ is continuous at $\theta = 1$, but suffers a jump in slope J given by equation (5.5.4). The vertical displacement for beryllium continues to increase, the curve being smooth at $\theta = T_+$, to its maximum value before slowly relaxing to the static value $w_0(\infty)$ given by equation (5.5.5). The graph of $w_0(\theta)$ for beryllium is very similar to that shown for magnesium in Figure 72, and only confirms the observation that beryllium really belongs to the (1) class of materials of Table 11.

Apatite, cadmium and zinc have reciprocal square root singularities at $\theta = T_+$ $(\tau = hT_+)$ caused by the conical point on the wave front. The vertical displacement for apatite and cadmium monotonically decreases from zero at $\theta = 1$ to $-\infty$ at $\theta = T_+$. The displacement singularity at $\theta = T_+$ is one sided, since the value of $w_0(T_+ + 0)$ is finite and positive for these two crystals. Their displacement curves then further increase to their maximum values before relaxing to the static value $w_0(\infty)$.

The epicenter motion of zinc is more complicated and the graph for this material masks some of the more interesting features, due to the compressed interval over which they occur. The $w_0(\theta)$ function for zinc decreases from zero at $\theta = 1$ to a local minimum value of -14.48828 at $\theta = 1.126905$. The graph then increases to zero at the θ value for which $f(\omega_3) = 0$, i.e.,

$$
\theta_0 = \left[\frac{(\alpha + 1 - \gamma/\beta)}{\alpha} \right]^{1/2},
\tag{5.5.6}
$$

or $\theta_0 = 1.127101$. The behaviour of the displacement is most interesting at this point. By appropriately expanding $w_0(\theta)$, as given in (5.4.38), about this zero it is found that for small $|\theta - \theta_0|$,

$$
w_0(\theta) \approx
\begin{cases}
-100099.23 (\theta - \theta_0) & \text{for} \quad \theta \geqslant \theta_0, \\
-2533.92\sqrt{\theta_0 - \theta} & \text{for} \quad \theta \leqslant \theta_0.
\end{cases}
\tag{5.5.7}
$$

Thus, $w_0(\theta)$ is continuous but nondifferentiable (hence causing a strain singularity) at this point. This kink in the displacement curve, which is caused by a head wave contribution, will be discussed more fully in a somewhat different context in section 5.6 below. Beyond this zero the displacement falls to its singular value of $-\infty$ at $\theta = T_+$. Now, however, the singularity is experienced on both sides of $\theta = T_+$ but the asymptotic form has different multiplicative constants on the two sides,

$$
w_0(\theta) \sim
\begin{cases}
-0.3906 (T_+ - \theta)^{-1/2} & \text{for} \quad \theta < T_+, \\
-0.4509 (\theta - T_+)^{-1/2} & \text{for} \quad \theta > T_+.
\end{cases}
\tag{5.5.8}
$$

The graph of $w_0(\theta)$ for zinc then increases to zero at $\theta = 1.132041$ and then continues to increase to its maximum value of 2.565049 at $\theta = 2.79$ before relaxing slowly to its static value $w_0(\infty)$ given by (5.5.5). A graphic 'blow-up' of a tiny interval of the θ-axis for zinc is shown in Figure 74. The numerical features for the graphs of apatite, beryllium, cadmium and zinc are summarized in Table 16.

Most of the material presented in this and the preceeding sections of this chapter, including tables and figures, can be found in Payton [69] and [70].

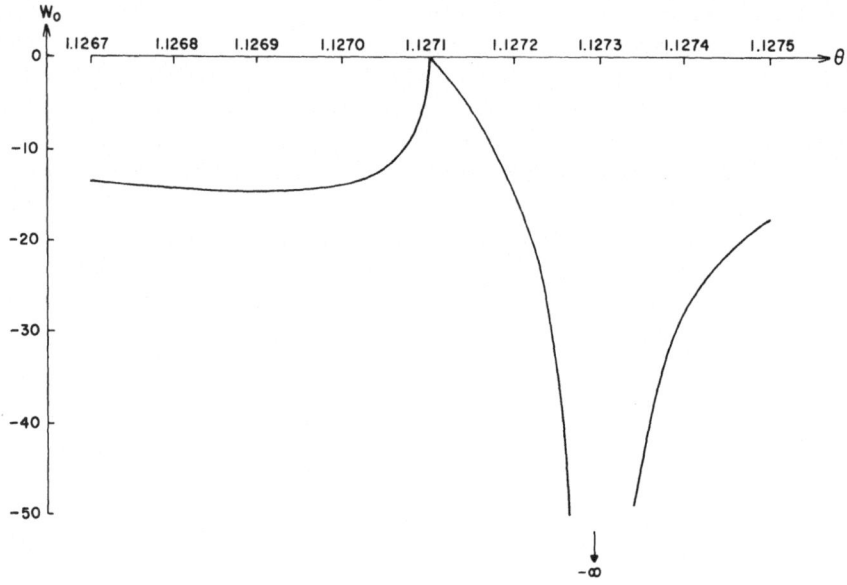

Figure 74. Vertical displacement for Zinc near $\theta = T_+$.

Table 16. Numerical features of $w_0(\theta)$ for four crystals

Crystal	$\alpha^{-1/2}$	$w_0(\alpha^{-1/2} + 0)$	T_+	$-J$	$w_0(T_+ + 0)$
Apatite	0.688428	0.436389	1.038088	12.243512	1.873675
Beryllium	0.695048	0.957066	1.099683	11.173724	2.692936
Cadmium	0.617802	0.283986	1.000277	1631.827243	1.841413
Zinc	0.798087	0.199409	1.127293	2.679413	$-\infty$

		Maxima		Minima		
Crystal	Zeros	θ	w_0	θ	w_0	$w_0(\infty)$
Apatite	1	0.778	0.450635			2.606024
		2.33	2.646827			
Beryllium		5.91	2.995686			2.994975
Cadmium	1	0.983	0.713003			2.329108
		2.58	2.356219			
Zinc	1	1.127	0	1.126905	-14.48828	2.522552
	1.127101	2.79	2.565049			
	1.132041					

6. Vertical Displacement Along the Epicentral-Axis Due to a Surface Point Load Acting Normal to the Surface

Up to the present point, the analysis of this chapter may seem to the reader to be rather narrow in scope since the displacement expressions have been obtained only at one point in space. However it will now be shown, by using a dynamic version of the Betti-Rayleigh reciprocal theorem (for a proof of the static theorem see e.g. Sokolnikoff [3]), that the epicenter displacement due to a buried point source can readily be used to find the epicentral-axis motion caused by an applied normal surface point load. The dynamic reciprocal theorem was first given by Graffi [71] for an isotropic body, while Knopoff and Gangi [72] stated the theorem for an anisotropic solid. Wheeler [73] has given a proof of the theorem which is rigorous for an unbounded anisotropic solid. A proof of the reciprocal theorem will now be sketched, without due regard for rigor. The proof here follows Graffi in that the time dependence is removed by means of the Laplace transform.

Denote the Laplace transform, with respect to time t, by an over bar, thus

$$\overline{(} = \int_0^\infty () e^{-st} dt. \tag{5.6.1}$$

Consider an anisotropic elastic solid, initially at rest, with volume and surface domains V and S. Two problems are posed, an unprimed problem and a primed problem, the displacement fields for which are caused by different body and surface forces in the two cases. Both sets of displacements must satisfy the basic equations of section 1.1. In particular

$$\bar{\sigma}_{ij,j} + \rho \bar{f}_i = \rho s^2 \bar{u}_i \quad \text{in} \quad V,$$

$$\bar{T}_i \equiv \bar{\sigma}_{ij} n_j = \bar{P}_i \quad \text{on} \quad S, \tag{5.6.2}$$

$$\bar{\sigma}_{ij} = c_{ijkl} \bar{u}_{k,l} \quad \text{in} \quad V,$$

and a corresponding primed problem

$$\bar{\sigma}'_{ij,j} + \rho \bar{f}'_i = \rho s^2 \bar{u}'_i \quad \text{in} \quad V,$$

$$\bar{T}'_i \equiv \bar{\sigma}'_{ij} n_j = \bar{P}'_i \quad \text{on} \quad S, \tag{5.6.3}$$

$$\bar{\sigma}'_{ij} = c_{ijkl} \bar{u}'_{k,l} \quad \text{in} \quad V.$$

In (5.6.2) and (5.6.3) $\bar{f}_i$, $\bar{P}_i$, $\bar{f}'_i$ and $\bar{P}'_i$ are the (transformed) applied forces. Consider next the integrals

$$\int_V \rho \bar{f}_i \bar{u}'_i dV + \int_S \bar{T}_i \bar{u}'_i dS \tag{5.6.4}$$

$$= \int_V (\rho s^2 \bar{u}_i - \bar{\sigma}_{ij,j}) \bar{u}'_i dV + \int_S \bar{\sigma}_{ij} \bar{u}'_i n_j dS \quad \text{using (5.6.2)}$$

$$= \int_V (\rho s^2 \bar{u}_i - \bar{\sigma}_{ij,j}) \bar{u}'_i dV + \int_V (\bar{\sigma}_{ij} \bar{u}'_i)_{,j} dV$$

$$= \int_V (\rho s^2 \bar{u}_i \bar{u}'_i + \bar{\sigma}_{ij} \bar{u}'_{i,j}) dV. \tag{5.6.5}$$

179

But $\bar{\sigma}_{ij}\bar{u}'_{i,j} = c_{ijkl}\bar{u}_{k,l}\bar{u}'_{i,j}$ let $(i,j,k,l) \rightarrow (k,l,i,j)$

$$= c_{klij}\bar{u}_{i,j}\bar{u}'_{k,l}$$

$$= c_{ijkl}\bar{u}'_{k,l}\bar{u}_{i,j} \quad \text{from } (1.1.5)$$

$$= \bar{\sigma}'_{ij}\bar{u}_{i,j},$$

so that

$$\bar{\sigma}_{ij}\bar{u}'_{i,j} = \bar{\sigma}'_{ij}\bar{u}_{i,j}. \tag{5.6.6}$$

Then from (5.6.5)

$$\int_V (\rho s^2 \bar{u}_i \bar{u}'_i + \bar{\sigma}_{ij}\bar{u}'_{i,j})dV$$

$$= \int_V (\rho s^2 \bar{u}_i \bar{u}'_i + \bar{\sigma}'_{ij}u_{i,j})dV$$

$$= \int_V \{\rho s^2 \bar{u}_i \bar{u}'_i - \bar{\sigma}'_{ij,j}\bar{u}_i + (\bar{\sigma}'_{ij}\bar{u}_i)_{,j}\} \, dV$$

$$= \int_V (\rho s^2 \bar{u}'_i - \bar{\sigma}'_{ij,j})\bar{u}_i dV + \int_S \bar{\sigma}'_{ij}\bar{u}_i n_j dS \quad \text{then use } (5.6.3)$$

$$= \int_V \rho \bar{f}'_i \bar{u}_i dV + \int_S \bar{T}'_i \bar{u}_i \, dS. \tag{5.6.7}$$

Hence from (5.6.4) and (5.6.7)

$$\int_V \rho \bar{f}_i \bar{u}'_i dV + \int_S \bar{T}_i \bar{u}'_i dS = \int_V \rho \bar{f}'_i \bar{u}_i dV + \int_S \bar{T}'_i \bar{u}_i dS, \tag{5.6.8}$$

which is the desired version of the reciprocal theorem for an (initially at rest) anisotropic solid. Using (5.6.2) and (5.6.3), (5.6.8) may also be written as

$$\int_V \rho \bar{f}_i \bar{u}'_i dV + \int_S \bar{P}_i \bar{u}'_i dS = \int_V \rho \bar{f}'_i \bar{u}_i dV + \int_S \bar{P}'_i \bar{u}_i dS. \tag{5.6.9}$$

Now let problem *I* (the primed problem) be the response of a transversely isotropic elastic half-space due to a buried point load, then in vector notation

$$\mathbf{f}'(\mathbf{x}, \mathbf{x}_0, t) = (a_1, a_2, a_3)\delta(x - x_0)\delta(y - y_0)\delta(z - z_0)H(t), \tag{5.6.10}$$

with stress-free surface

$$\bar{\mathbf{T}}' = 0 \quad \text{on} \quad z = 0. \tag{5.6.11}$$

Denote the solution to problem I by

$$\mathbf{u}'(\mathbf{x}, \mathbf{x}_0, t) \equiv (a_1 u'_1 + a_2 u'_2 + a_3 u'_3, a_1 v'_1 + a_2 v'_2 + a_3 v'_3, a_1 w'_1$$

$$+ a_2 w'_2 + a_3 w'_3). \tag{5.6.12}$$

Let problem II (the unprimed problem) be the response of a transversely isotropic elastic half-space with no body force,

$$\mathbf{f}(\mathbf{x}, t) = 0,$$ (5.6.13)

subjected to a point surface load

$$\mathbf{T} = (b_1, b_2, b_3)\delta(x - x_1)\delta(y - y_1)H(t).$$ (5.6.14)

Denote the solution to problem II by

$$\mathbf{u}(\mathbf{x}, t) = (u, v, w).$$ (5.6.15)

Invoking the reciprocal theorem (5.6.8) and using the fact that a_1, a_2 and a_3 are independent constants gives

$$\rho u(\mathbf{x}_0, t) = (b_1 u_1' + b_2 v_1' + b_3 w_1', b_1 u_2' + b_2 v_2' + b_3 w_3', b_1 u_3'$$
$$+ b_2 v_3' + b_3 w_3'),$$ (5.6.16)

where the argument in each displacement function of the right side of (5.6.17) is $(x_1, y_1, 0; x_0, y_0, z_0, t)$. Specialize now so that the buried source is applied on the z-axis ($x_0 = y_0 = 0$), the surface load is applied at the origin ($x_1 = y_1 = 0$) and its direction is normal to the surface ($b_1, b_2 = 0$). Then, from symmetry

$$w_1'(0, 0, 0; 0, 0, z_0, t) = 0 \text{ and } w_2'(0, 0, 0; 0, 0, z_0, t) = 0,$$ (5.6.17)

so that from (5.6.16)

$$u(0, 0, z_0, t) = 0, v(0, 0, z_0, t) = 0 \quad \text{and}$$

$$w(0, 0, z_0, t) = \frac{b_3}{\rho} w_3'(0, 0, 0; 0, 0, z_0, t).$$ (5.6.18)

Here $a_3 w_3'(0, 0, 0; 0, 0, z_0, t)$ is the vertical component of displacement at the epicenter due to a vertical point force

$$\rho \mathbf{f} = (0, 0, a_3)\rho\delta(x)\delta(y)\delta(z - z_0)H(t),$$ (5.6.19)

acting at a depth $z_0 > 0$ below the surface and on the z-axis.

Thus, upon adapting the notation of the present chapter, and replacing z_0 by z, the epicentral-axis displacement caused by a normal surface point load of strength b_3 is

$$u(0, 0, z, t) = 0, \qquad v(0, 0, z, t) = 0 \quad \text{and}$$
$$w(0, 0, z, t) = \frac{b_3}{4\pi c_{44} z} w_0(\theta),$$ (5.6.20)

where now $\theta = \tau/z$. Hence the results found in sections 5.2 and 5.4 for the epicenter motion caused by a buried source, may be used to find the epicentral-axis motion due to a surface point force. A similar use of the reciprocal theorem for an isotropic body has been made by Pekeris and, Lifson [74].

An observer situated on the z-axis beneath the point of application of the surface point load, will experience four distinct wave arrivals in the case of zinc. These are (i) $\theta = \alpha^{-1/2}$ due to the outer sheet of the wave front, (ii) $\theta = 1$ due to the base of the zinc wave front cone, (iii) $\theta = \theta_0$ and (iv) $\theta = T_+$ due to the conical point on the wave front. It should be noted that the arrivals $\theta = \alpha^{-1/2}$, 1 and T_+ will all exist for the corresponding free-space problem but that the arrival at $\theta = \theta_0$ will be absent in this case. It is therefore apparent that the arrival at $\theta = \theta_0$ is somehow connected with the fact that a bounding surface is present. A lucid explanation of this point on the wave front has been given by Musgrave and Payton [75]. This explanation will now be briefly summarized.

Based on the discussion of sections 5.3 and 5.4, necessary conditions for the existence of a wave front at $\theta = \theta_0$ (for the point surface loaded half-space problem under consideration) are

$$\gamma < (\beta + 1), \tag{5.6.21}$$

since then the outer branch of the slowness curve has a local minimum on the symmetry axis, and

$$-1/\beta > \xi_+ \tag{5.6.22}$$

which (from Figure 70) implies that an additional branch point at $\omega = -1/\beta$ is encountered on the Cagniard integration path. Both of these conditions are satisfied for zinc but offer little insight towards a complementary physical interpretation of the phenomenon.

When a half-space $z > 0$ is loaded on its surface by a point source, energy is fed into the interior by rays emanating directly from the source and also by rays that first move along the surface and radiate into the interior. The envelope of the latter set of wavelets is known as a head wave. It was shown in [75] that the image of the head wave front (relevant to the present discussion) on the slowness diagram section is a line from the point $(0, \beta^{-1/2})$ in the (p, q) plane, tangent to the inner branch of the slowness curve (hence parallel to the p-axis) and terminating at the point $(p_A, \beta^{-1/2})$ on the outer branch of the slowness curve. Broadly speaking, there are three possibilities (finer subdivisions are explored in [75]),

$$\beta^{-1/2} < Q, \tag{5.6.23}$$

$$\beta^{-1/2} = Q, \tag{5.6.24}$$

and

$$\beta^{-1/2} > Q, \tag{5.6.25}$$

where the point (P, Q) has a horizontal tangent (see figure 71). If $\beta^{-1/2} < Q$, the head wave front crosses the z-axis, forming an additional conical point at the instant

$$\tau = P_A z. \tag{5.6.26}$$

But this is precisely the situation for zinc, since by direct calculation

$$Q^2 = -\xi_+ \tag{5.6.27}$$

hence $-1/\beta > \xi_+$ from (5.6.23) in agreement with condition (5.6.22), and

$$p_A = [(\alpha + 1 - \gamma/\beta)/\alpha]^{1/2} \tag{5.6.28}$$

so that from (5.5.6) $\theta = \theta_0$, having used $\theta = \tau/z$ here.

The properties of head waves in a half-space of arbitrary anisotropy have featured in the work of Willis and Bedding [76, 77] and more recently in a paper by Musgrave and Payton [78]. The special nature of the situation (5.6.24), when the arrival times of the conical points on the wave surface and on the head wave front coincide, is treated in Payton and Musgrave [79].

7. Body Forces Equivalent to Internal Discontinuities in an Anisotropic Elastic Solid

Before closing this chapter (and monograph) one further application of the reciprocal theorem (5.6.8) will be given. Let the bounding surface S for problem (5.6.2) be composed of an outer surface S_1 where homogeneous boundary conditions apply, plus an internal dislocation (or fault) surface Σ across which the displacement vector and traction vector may experience jumps denoted by $[u_i(x, t)]$ and $[T_i(x, t)]$. Here the position vector x lies on the dislocation Σ. For the primed problem let

$$\rho \bar{f}'_i = \delta_{ip} \delta(x - \xi), \tag{5.7.1}$$

where $\delta(x - \xi) \equiv \delta(x_1 - \xi_1) \delta(x_2 - \xi_2) \delta(x_3 - \xi_3)$.

Let the primed problem have the same homogeneous boundary conditions on S_1, but have continuous displacement and traction vectors on the cut surface Σ. Denote the solution of the primed problem by

$$\bar{u}'_i \equiv \bar{G}'_{ip}(x|\xi, s). \tag{5.7.2}$$

Then from (5.6.8)

$$\int_V \rho \bar{f}_i(x, s) \bar{G}'_{ip}(x|\xi\, s) dV_x - \int_\Sigma [\bar{T}_i(\eta, s)] \bar{G}'_{ip}(\eta|\xi, s) d\Sigma_\eta$$

$$= \bar{u}_p(\xi, s) - \int_\Sigma c_{ijkl} \bar{G}'_{kp,l}(\eta|\xi, s) n_j [\bar{u}_j(\eta, s)] d\Sigma_\eta. \tag{5.7.3}$$

In the second integral on the right side of (5.7.3) use has been made of the relation

$$\bar{T}'_i = \bar{\sigma}'_{ij} n_j = c_{ijkl} \bar{G}'_{kp,l}(x|\xi, s) n_j, \tag{5.7.4}$$

where n_j is the unit (outward drawn) normal vector on Σ_η, $x = \eta$ when x is on Σ and

$$(\ )_{,l} \text{ means } \frac{\partial(\ )}{\partial x_l}. \tag{5.7.5}$$

In equation (5.7.3), the negative sign in front of the two dislocation integrals stems from the integration over both sides of Σ. Also note that the subscript 'x' on the differentials in (5.7.3) is to remind the reader that the integration is in 'x' space. From symmetry

$$\bar{G}'_{ip}(x|\xi, s) = \bar{G}'_{pi}(\xi|x, s) \tag{5.7.6}$$

so that (5.7.3) may be written

$$\bar{u}_p(\xi, s) = \int_V \bar{G}'_{pi}(\xi|x, s) \rho \bar{f}_i(x, s) dV_x$$

$$+ \int_\Sigma n_j [\bar{u}_i(\boldsymbol{\eta}, s)] \, c_{ijkl} \bar{G}'_{pk,l}(\boldsymbol{\xi}|\boldsymbol{\eta}, s) d\Sigma_\eta$$

$$- \int_\Sigma [\bar{T}_i(\boldsymbol{\eta}, s)] \, \bar{G}'_{pi} (\boldsymbol{\xi}|\boldsymbol{\eta}, s) \, d\Sigma_\eta. \tag{5.7.7}$$

By the sifting property of the Dirac delta, $\bar{G}'_{pi}$ and $\bar{G}'_{pk,l}$ may be written as x-space volume integrals

$$\bar{G}'_{pi}(\boldsymbol{\xi}|\boldsymbol{\eta}, s) = \int_V \bar{G}'_{pi}(\boldsymbol{\xi}|\mathbf{x}, s) \delta(\mathbf{x} - \boldsymbol{\eta}) dV_x, \tag{5.7.8}$$

and

$$-\bar{G}'_{pk,l}(\boldsymbol{\xi}|\boldsymbol{\eta}, s) = \int_V \bar{G}'_{pk}(\boldsymbol{\xi}|\mathbf{x}, s) \frac{\partial}{\partial x_l} \delta(\mathbf{x} - \boldsymbol{\eta}) dV_x. \tag{5.7.9}$$

Then equation (5.7.7) may be written as

$$\bar{u}_p(\boldsymbol{\xi}, s) = \int_V \bar{G}'_{pk}(\boldsymbol{\xi}|\mathbf{x}, s) \{ \rho \bar{f}_k (\mathbf{x}, s) - \int_\Sigma [\bar{T}_k(\boldsymbol{\eta}, s)] \, \delta(\mathbf{x} - \boldsymbol{\eta}) d\Sigma_\eta$$

$$- \int_\Sigma n_j [\bar{u}_i(\boldsymbol{\eta}, s)] \, c_{ijkl} \frac{\partial}{\partial x_l} \delta(\mathbf{x} - \boldsymbol{\eta}) d\Sigma_\eta \} dV_x. \tag{5.7.10}$$

Call

$$\rho \bar{e}_k(\mathbf{x}, s) = - \int_\Sigma \{ n_j [\bar{u}_i(\boldsymbol{\eta}, s)] \, c_{ijkl} \frac{\partial}{\partial x_l} \delta (\mathbf{x} - \boldsymbol{\eta})$$

$$+ [\bar{T}_k(\boldsymbol{\eta}, s)] \, \delta(\mathbf{x} - \boldsymbol{\eta}) \} d\Sigma_\eta. \tag{5.7.11}$$

Then, upon returning to the time domain, the body forces equivalent to the dislocation sources $[u_i(\boldsymbol{\eta}, t)]$ and $[\bar{T}_k(\boldsymbol{\eta}, t]$ are

$$\rho e_k(\mathbf{x}, t) = - \int_\Sigma \{ n_j [u_i(\boldsymbol{\eta}, t)] \, c_{ijkl} \frac{\partial}{\partial x_l} \delta(\mathbf{x} - \boldsymbol{\eta})$$

$$+ [T_k(\boldsymbol{\eta}, t)] \, \delta(\mathbf{x} - \boldsymbol{\eta}) \} d\Sigma_\eta. \tag{5.7.12}$$

Equation (5.7.12) is the celebrated formula of Burridge and Knopoff [68] which has been extensively used as a model for earthquake source dislocations for an isotropic solid. A more careful derivation of (5.7.12) (as given in [68]) would reveal that this formula remains valid for a non-homogeneous elastic body also. According to Burridge and Knopoff [68], when the body force (5.7.12) is applied in the absence of a dislocation, a radiation pattern is produced identical to that of the dislocation.

The subject of time dependent (e.g. moving) dislocations in anisotropic media has received little attention insofar as explicit solutions are concerned. The general theory of such problems has been considered by several authors, including Mura [80]. The Green tensors obtained in Chapters 3 and 5 of this monograph may help in the solution of such problems. Recent research connected with dislocations in crystals by Kosevich [81] and Steeds and Willis [82] as well as the importance of anisotropy in seismology (see Crampin [83] and related papers in this reference) point to the need for further research in this direction.

References

[1] Green, A.E. and Zerna, W., 'Theoretical Elasticity.' London, Oxford University Press, 1954.

[2] Pearson, C.E., 'Theoretical Elasticity.' Cambridge Massachusetts, Harvard University Press, 1959.

[3] Sokolnikoff, I.S., 'Mathematical Theory of Elasticity.' New York, McGraw-Hill Book Company (second edition), 1956.

[4] Fedorov, F.I., 'Theory of Elastic Waves in Crystals.' New York, Plenum Press, 1968.

[5] Landolt-Börnstein, 'Numerical Data and Functional Relationships in Science and Technology.' New Series, Group 3, vol. I, article by R.F.S. Hearmon. Berlin-Heidelberg-New York, Springer, 1969.

[6] Noble, B., 'Applied Linear Algebra.' Englewood Cliffs, New Jersey, Prentice-Hall, 1969.

[7] John, F., 'Plane Waves and Spherical Means Applied to Partial Differential Equations.' New York, John Wiley and Sons, 1955.

[8] Gel'fand, I.M. and Shilov, G.E., 'Generalized Functions,' vol. I, New York, Academic Press, 1964.

[9] Truesdell, C. and Noll, W., 'The Non-Linear Field Theories of Mechanics.' Berlin-Heidelberg-New York, Springer, 1965.

[10] Knops, R.J. and Payne, L.E., 'Uniqueness Theorems in Linear Elasticity.' Berlin-Heidelberg-New York, Springer, 1971.

[11] Courant, R. and Hilbert, D., 'Methods of Mathematical Physics,' vol. I. New York, Interscience Publishers, Inc., 1953.

[12] Payton, R.G., Int. J. Engng. Sci. *13,* 183 (1975).

[13] Payton, R.G., Arch. Rational Mech. Anal. *32,* 311 (1969).

[14] Payton, R.G., Instituto Lombardo (Rend. Sc.) A *108,* 684 (1974).

[15] Lighthill, M.J., Phil. Trans. Roy. Soc. London A *252,* 397 (1960).

[16] Born, M. and Wolf, E., 'Principles of Optics.' London, Pergamon Press, 1959.

[17] Broutman and Krock (editors) 'Treatise of Composite Materials' vol. 9, Chamis, C.C. (editor) 'Structural Analysis and Design,' article by Moon, F.C. New York, Academic Press, 197 .

[18] Courant, R. and Hilbert, D., 'Methods of Mathematical Physics,' vol. II. New York, Interscience Publishers, Inc., 1962.

[19] Kline, M. and Kay, I.W., 'Electromagnetic Theory and Geometric Optics.' New York, Interscience Publishers, Inc., 1965.

[20] Musgrave, M.J.P., 'Crystal Acoustics.' San Francisco, Holden-Day, Inc., 1970.

[21] Duff, G.F.D., Phil. Trans. Roy. Soc. London A *252,* 31 (1960).

[22] Salmon, G., 'Higher Plane Curves.' New York, Chelsea Publishing Co.

[23] Payton, R.G., Arch. Rational Mech. Anal. *35,* 402 (1969).

[24] Garnir, H. G. (editor) 'Boundary Value Problems for Linear Evolution Partial Differential Equations,' article by G.F.D. Duff. Dordrecht, Holland, D. Reidel Publishing Co., 1977.

[25] Thom, R., 'Structural Stability and Morphogenesis.' Reading, Massachusetts, Benjamin, 1975.

[26] Santalo, L. A., 'Integral Geometry and Geometric Probability.' Reading, Massachusetts, Addison-Wesley, 1976.

[27] Weitzner, H., Phys. Fluids *4,* 1238 (1961).

[28] Bazer, J. and Yen, D.H.Y., Comm. Pure Appl. Math. *22,* 279 (1969).

[29] Payton, R.G., J. Math. Anal. Appl. *54,* 200 (1976).

[30] Burridge, R., Q. J. Mech. Appl. Math. *20,* 41 (1966).

[31] Burridge, R., Proc. Cambridge Philos. Soc. *63* (1967), 819.

[32] Petrowski, I.G., Rec. Math. [Mat. Sbornik] *17* (1945), 289.

[33] Payton, R.G., Proc. Camb. Phil. Soc. *70* (1971), 191.

[34] Payton, R.G., ZAMP *29* (1978), 262.

[35] Budaev, V.S., Prikl. Mekhan. *9* (1973), 67.

[36] Kraut, E.A., Ph.D. thesis, University of California, Los Angeles (1962).

[37] Bazer, J. and Yen, D.H.Y., Comm. Pure Appl. Math. *20* (1967), 329.

[38] Ludwig, D., Pacific J. Math. *15* (1965), 215.

[39] Willis, J.R., Phil. Trans. Roy. Soc. A *274* (1973), 435.

[40] Stoker, J.J., 'Differential Geometry.' New York, John Wiley and Sons, 1969.

[41] Payton, R.G., Proc. Camb. Phil. Soc. *72* (1972), 105.

[42] Payton, R.G., Q. Jl. Mech. Appl. Math. *28* (1975), 473.

[43] Carrier, G.F., Q. Appl. Math. *4* (1946), 160.

[44] Cameron, N. and Eason, G., Q. Jl. Mech. Appl. Math. *20* (1967), 23.

[45] Stokes, G.G., Trans. Camb. Phil. Soc. *9* (1851), 1.

[46] Poisson, S.D., Mem. Acad. Sci. Paris *8* (1829), 623.

[47] Cagniard, L., 'Reflection and Refraction of Progressive Seismic Waves.' New York, McGraw-Hill Book Company (1962).

[48] Pekeris, C.L., Proc. natn. Acad. Sci. *41* (1955), 629.

[49] Payton, R.G., Q. Appl. Math. *35* (1977), 63.

[50] Chee-Seng, L., Proc. Camb. Philos. Soc. *74* (1973), 369.

[51] Buchwald, V.T., Proc. Roy. Soc. London A *253* (1959), 563.

[52] Herglotz, G., Berlin. Verhandl. Sächs. Acad. Wiss. Leipzig, Math. Phys. Kl. *78* (1926), 93; *80* (1928), 60.

[53] Burridge, R., Q. Jl. Mech. Appl. Math. *20* (1967), 41.

[54] Barry, P.A. and Musgrave, M.J.P., Q. Jl. Mech. Appl. Math. *32* (1979), 205.

[55] Gel'fand, I.M., Graev, M.I. and Vilenkin, N.Ya., 'Generalized Functions,' vol. V. New York, Academic Press, 1966.

[56] Burridge, R., Proc. Roy. Soc. A *276* (1963), 367.

[57] Lamb, H., Phil. Trans. Roy. Soc. A *203* (1904), 1.

[58] Kraut, E.A., Rev. Geophys. *1* (1963), 401.

[59] Burridge, R., Q. Jl. Mech. Appl. Math. *24* (1971), 81.

[60] Budaev, V.S., Zh. Prikl. Mekh. Tekh. Fiz. no. *3* (1974), 121.

[61] Widder, D.V., Trans. Amer. Math. Soc. *36* (1934), 107.

[62] Stonely, R., Mon. Not. R. astr. Soc. geophys. Suppl. *5* (1949), 343.

[63] Aki, K. and Richards, P.G., 'Quantitative Seismology, Theory and Methods.' San Francisco, W.H. Freeman and Company, 1980.

[64] Phinney, R.A., J. Geophy. Res. *66* (1961), 1445.

[65] Gilbert, F. and Laster, S.J., Bull. Seis. Soc. Amer. *52* (1962), 59.

[66] Chapman, C.H., Pure and Appl. Geophys. *94* (1972), 233.

[67] Ryan, R.L., J. Sound Vib. *14* (1971), 511.

[68] Burridge, R. and Knopoff, L., Bull. Seis. Soc. Amer. *54* (1964), 1875.

[69] Payton, R.G., Int. J. Engng. Sci. *17* (1979), 879.

[70] Payton, R.G., SIAM J. Appl. Math. *40* (1981), 373.

[71] Graffi, D., Mem. Accad. Sci. Bologna *4* (1946/1947), 103.

[72] Knopoff, L. and Gangi, A.F., Geophy. *24* (1959), 681.

[73] Wheeler, L.T., Q. Appl. Math. *28* (1970), 91.

[74] Pekeris, C.L. and Lifson, H., J. Acoust. Soc. Amer. *29* (1957), 1233.

[75] Musgrave, M.J.P. and Payton, R.G., Q. Jl. Mech. Appl. Math. *34* (1981), 235.

[76] Willis, J.R. and Bedding, R.J., Math. Proc. Camb. Phil. Soc. *77* (1975), 591.

[77] Willis, J.R. and Bedding, R.J., 'Modern Problems in Elastic Wave Propagation' (Eds. J.D. Achenbach and J. Miklowitz). New York, John Wiley and Sons (1978), 347.

[78] Musgrave, M.J.P. and Payton, R.G., Q. Jl. Mech. Appl. Math. *35* (1982), 173.

[79] Payton, R.G. and Musgrave, M.J.P., J. Elasticity *13* (1983), 149.

[80] Mura, T. (editor) 'Mathematical Theory of Dislocations.' The American Society of Mechanical Engineers, New York (1969), 25.

[81] Kosevich, A.M., 'Dislocations in Solids' vol. 1, (F.R.N. Nabarro, editor) North-Holland Publishing Co., Amsterdam (1979), 33.

[82] Steeds, J.W. and Willis, J.R., 'Dislocations in Solids' vol. 1, (F.R.N. Nabarro, editor) North-Holland Publishing Co., Amsterdam (1979), 143.

[83] Crampin, S., Geophy. J. Roy. Astron. Soc. *49* (1977), 9.

[84] Osipov, I.O., J. Appl. Math. and Mech., *33* (1969) 534.

[85] Musgrave, M.J.P. and Payton, R.G., J. Elasticity (not yet published).

[86] Musgrave, M.J.P., Proc. Camb. Phil. Soc., *53* (1957) 897.

[87] Achenbach, J.D., 'Wave Propagation in Elastic Solids.' Amsterdam, North-Holland Publishing Co. 1973.

[88] Eringen, A.C. and Suhubi, E.S., 'Elastodynamics' vol. II. New York, Academic Press 1975.

[89] Miklowitz, J., 'Elastic Waves and Waveguides.' Amsterdam, North-Holland Publishing Co. 1978.

[90] Hudson, J.A., 'The Excitation and Propagation of Elastic Waves.' Cambridge, Cambridge University Press 1980.

Author index

Author index

Subject index